Gen Z, Tourism, and Sustainal Consumption

Gen Z, Tourism, and Sustainable Consumption is the first book to provide a comprehensive account of Generation Z in relation to sustainable consumption practices and travel cultures.

Gen Z is regarded as the world's largest consumer market. The growth and behaviour of this economically significant market will have enormous implications for the future development of the tourism industry and destinations and its long-term sustainability. Characterised as being the first generation to grow up with the Internet and sometimes even referred to as the i-Generation, Gen Z is broadly regarded as having an avid interest in travel but seeks to do so in a way that is socially and environmentally conscious, mobile connected, and grounded in authentic local experiences. Logically structured and featuring contributions from a plethora of experts on the topic, this volume provides a critical examination of Gen Z consumer and travel behaviour in a comparative international context and its implications for the tourism, hospitality, and events industries.

Embellished with illustrative figures and tables throughout, this book will be of pivotal interest not only to policy makers, destination management and marketing organisations (DMOs), and students of tourism, hospitality, sustainable consumption, and consumer culture, but also to those who seek to cater to this key market.

Siamak Seyfi is an Assistant Professor at the Geography Research Unit of the University of Oulu, Finland. He is also an Adjunct Professor at the Department of Tourism Marketing of the University of Eastern Finland, and a Visiting Associate Professor at the School of Hospitality, Tourism and Events, Taylor's University, Malaysia. Using a multidisciplinary/interdisciplinary approach and informed by diverse disciplinary perspectives, his research focuses on tourism mobilities, tourist behaviour, resilience, sustainability, and politics of tourism and peace through tourism. Guided by generational and lifestyle theories, his recent research focuses on political and ethical consumerism with a special focus on the Gen Z cohort.

C. Michael Hall is Ahurei Professor in the Department of Management, Marketing and Tourism, University of Canterbury, New Zealand; Visiting Professor and Docent in Geography, University of Oulu, Finland; Visiting Professor, School of Business and Economics, Linnaeus University, Kalmar, Sweden; Guest Professor, Department of Service Management and Service Studies, Lund University, Helsingborg, Sweden; Visiting Professor, CRiC, Taylor's University, Kuala Lumpur, Malaysia; and Eminent Scholar, Kyung Hee University, Seoul, South Korea. Co-editor of *Current Issues in Tourism* and Field Editor of *Frontiers in Sustainable Tourism*, he publishes widely on tourism, sustainability, global environmental change, food, and regional development.

Marianna Strzelecka is Associate Professor at the School of Business and Economics of Linnaeus University, Sweden. She conducts research at the intersection of tourism, rural sociology, and environmental social sciences. Her most recent projects deal with conservation conflicts, a sense of justice, and nature stewardship. While studying nature stewardship in volunteer travel, she has explored hedonic values, self-efficacy, and global citizenship attitudes.

Contemporary Geographies of Leisure, Tourism and Mobility
Series Editor: C. Michael Hall
Professor at the Department of Management, College of Business and Economics, University of Canterbury, Christchurch, New Zealand

The aim of this series is to explore and communicate the intersections and relationships between leisure, tourism and human mobility within the social sciences.

It will incorporate both traditional and new perspectives on leisure and tourism from contemporary geography, e.g. notions of identity, representation and culture, while also providing for perspectives from cognate areas such as anthropology, cultural studies, gastronomy and food studies, marketing, policy studies and political economy, regional and urban planning, and sociology, within the development of an integrated field of leisure and tourism studies.

Also, increasingly, tourism and leisure are regarded as steps in a continuum of human mobility. Inclusion of mobility in the series offers the prospect to examine the relationship between tourism and migration, the sojourner, educational travel, and second home and retirement travel phenomena.

The series comprises two strands:

Contemporary Geographies of Leisure, Tourism and Mobility aims to address the needs of students and academics, and the titles will be published simultaneously in hardback and paperback.

Routledge Studies in Contemporary Geographies of Leisure, Tourism and Mobility is a forum for innovative new research intended for research students and academics, and the titles will initially be available in hardback only. Titles include:

Inclusion in Tourism
Understanding Institutional Discrimination and Bias
Edited by Susan L. Slocum

Second Homes and Climate Change
Bailey Ashton Adie and C. Michael Hall

Gen Z, Tourism, and Sustainable Consumption
The Most Sustainable Generation Ever?
Edited by Siamak Seyfi, C. Michael Hall, and Marianna Strzelecka

For more information about this series, please visit: www.routledge.com/Contemporary-Geographies-of-Leisure-Tourism-and-Mobility/book-series/SE0522

Gen Z, Tourism, and Sustainable Consumption

The Most Sustainable Generation Ever?

Edited by Siamak Seyfi, C. Michael Hall, and Marianna Strzelecka

Routledge
Taylor & Francis Group

LONDON AND NEW YORK

First published 2024
by Routledge
4 Park Square, Milton Park, Abingdon, Oxon OX14 4RN

and by Routledge
605 Third Avenue, New York, NY 10158

Routledge is an imprint of the Taylor & Francis Group, an informa business

British Library Cataloguing-in-Publication Data
A catalogue record for this book is available from the British Library

ISBN: 978-1-032-26706-7 (hbk)
ISBN: 978-1-032-26707-4 (pbk)
ISBN: 978-1-003-28958-6 (ebk)

DOI: 10.4324/9781003289586

Typeset in Times New Roman
by Apex CoVantage, LLC

To Cooper, Michael's favourite Gen Z, may all your travels forever be happy and sustainable

Contents

Figures

Tables

Contributors

Fanny Aapio
University of Turku, Finland.

Marie-Louise Bank
Católica Lisbon School of Business & Economics, Palma de Cima, Portugal.

Monica Adele Breiby
Inland School of Business and Social Sciences, Inland Norway University of Applied Sciences, Norway.

Sandra Castro-González
Business Administration and Marketing Department, University of Santiago de Compostela, Spain.

Rui Augusto da Costa
Department of Economics, Management, Industrial Engineering and Tourism, University of Aveiro, Portugal.

Eva Duedahl
Center for Research on Digitalization and Sustainability, Inland Norway University of Applied Sciences, Norway.

Birgitta Ericsson
Eastern Norway Research Institute, Inland Norway University of Applied Sciences, Norway.

Fernando Almeida Garcia
Department of Geography, Faculty of Tourism, Malaga, Spain.

Abolfazl Siyamiyan Gorji
Department of Geography, Faculty of Tourism, Malaga, Spain.

Miia Grénman
Jyväskylä University School of Business and Economics, Finland.

C. Michael Hall
Department of Management, Marketing and Tourism, University of Canterbury, Christchurch, New Zealand; College of Hotel & Tourism Management, Kyung Hee University, Seoul, South Korea; Geography Research Unit, University of

Oulu, Oulu, Finland; School of Business and Economics, Linnaeus University, Kalmar, Sweden; Department of Service Management and Service Studies, Lund University, Helsingborg, Sweden; Centre for Research and Innovation in Tourism (CRiT), Taylor's University, Kuala Lumpur, Malaysia.

Cindy Yoonjoung Heo
EHL Hospitality Business School, HES-SO, University of Applied Sciences and Arts Western Switzerland.

Seyedasaad Hosseini
Department of Geography, Faculty of Tourism, Malaga, Spain.

Eugénie Le Bigot
PhD candidate, Laboratorie ESO-Caen, Université de Caen, France.

Merethe Lerfald
Eastern Norway Research Institute, Inland Norway University of Applied Sciences, Norway.

Rafael Cortes Macias
Department of Geography, Faculty of Tourism, Malaga, Spain.

Márcio Ribeiro Martins
Instituto Politécnico de Bragança, Portugal; CITUR (Centro de Investigação, Desenvolvimento e Inovação em Turismo); GOVCOPP (Research Group of Tourism and Development), University of Aveiro, Campus universitário de Santiago, Portugal.

Salvatore Monaco
Free University of Bozen-Bolzano, Italy.

Hogne Øian
Inland School of Business and Social Sciences, Inland Norway University of Applied Sciences, Norway.

Alicia Orea-Giner
Business Economics, Rey Juan Carlos University, Madrid, Spain.
EIREST, University of Paris 1 Panthéon-Sorbonne, Paris, France

Juulia Räikkönen
University of Turku, 20014 University of Turku, Finland.

Helena Maria Correia Neves Cordeiro Rodrigues
Instituto Universitário de Lisboa (ISCTE-IUL), Business Research Unit (BRU-IUL), Portugal.

Francisco Romera
Cornwall Business School, Falmouth University, United Kingdom.

Lionel Saul
EHL Hospitality Business School, HES-SO, University of Applied Sciences and Arts Western Switzerland.

Siamak Seyfi
Geography Research Unit, University of Oulu, Finland.
Centre for Research and Innovation in Tourism (CRiT), Taylor's University, Kuala Lumpur, Malaysia.

Marianna Strzelecka
School of Business and Economics, Linnaeus University, Kalmar, Sweden.

Jungho Suh
School of Social Sciences, The University of Adelaide, Australia.

Outi Uusitalo
Jyväskylä University School of Business and Economics, University of Jyväskylä, Finland.

Guadalupe Vila-Vazquez
Business Administration and Marketing Department, University of Santiago de Compostela, Spain.

Belén Bande Vilela
Business Administration and Marketing Department, University of Santiago de Compostela, Spain.

Acknowledgements

Nobody has been more important to Siamak in the pursuit of this project than the members of his family. He would like to thank his parents and especially his late father whose love and guidance are with him in everything he pursues. He also wishes to thank his loving and supportive wife, Mina, for her enduring love and patience.

Marianna would like to thank colleagues and friends for their support in her academic journey. She would like to especially thank B. Bynum Boley for continued support and inspiration over the years, Stephan Reinhold for challenging comments during work lunches, Solene Prince for her critical insights, and Per PPL for being himself. Deja DeMoss deserves special recognition for his support over the years and patience and Luna Strzelecka DeMoss for being the most inspiring human on Earth. Last, but not least, Marianna would like to express her gratitude to Jonas Ainouz and Katja Katja Törneman for the best climbing experience in Kalmar.

Michael would like to thank colleagues and friends with whom he has had relevant conversations or conducted research over the years in relation to this work, some of which are also contributors. In addition to thanking Siamak and Marianna for co-editing and those friends who have contributed to this book, Michael would like to thank Alberto Amore, Dorothee Bohn, Chris Chen, Tim Coles, Hervé Corvellec, David Duval, Martin Gren, Stephan Gössling, Peter Harinson, Johan Hultman, Tyron Love, Dieter Müller, Yael Ram, Jarkko Saarinen, Anna Dóra Sæþórsdóttir, Daniel Scott, Kimberley Wood, and Maria José Zapata-Campos for their thoughts on tourism and the world, as well as for the stimulation of Beirut, Paul Buchanan, Nick Cave, Bruce Cockburn, Elvis Costello, Stephen Cummings, David Bowie, Ebba Fosberg, Mark Hollis, Aimee Mann, Larkin Poe, Vinnie Reilly, Henry Rollins, Emma Swift, TISM, Henry Wagon, *The Guardian*, BBC6, JJ, and KCRW – for making the world much less confining. Special mention must also be given to Malmö Saluhall; Packhuset and Postgarten in Kalmar; and Hotel Lasaratti in Oulu. Finally, Michael would also like to extend his much belated but nevertheless heartfelt thanks and recognition to Jody Cowper-James for all her love and support for Michael's work over the years.

The editors would also like to gratefully acknowledge the assistance of Jody Cowper-James for her proofreading of much of the book. Finally, we would all like to thank our authors for dealing with COVID-19 and life-related delays to this book, and to Emma and all at Routledge for continuing to support this project despite the stresses of the last few years.

Part I
Introduction and context

1 Introduction

Gen Z, tourism, and sustainable consumption

Siamak Seyfi, C. Michael Hall,
and Marianna Strzelecka

Introduction: why does Gen Z matter as tourists?

Generational change has become increasingly important in influencing consumer behaviour, making it both a blessing and a curse for tourism destinations (Robinson & Schänzel, 2019). According to Yeoman et al. (2013), demographics and travel patterns specific to different generations are the key drivers of future tourism demand. Thus, understanding cross-generational traveller behaviours and how well the products and services of tourism providers cater to changing needs and expectations has been essential to tourism development and marketing plans (Haddouche & Salomone, 2018). Youth travel has evolved rapidly over the past few decades as a result of increased global mobility, technical advancements, and stronger physical, cultural, and political exchanges. This growing market, for example, was estimated to generate more than €400 billion in 2020 (Deloitte, 2020), making it an important segment for destinations worldwide. The unique character and economic significance of the young traveller have been noted in a UNWTO (United Nations World Tourism Organization) global report on the power of youth travel:

> Young people are more adventurous, looking for social contact with other young people and to discover new cultures and develop their knowledge. Because young people are inspired and motivated to travel as often as they possibly can, for longer periods of time and have an interest in visiting areas not frequented by traditional tourists
>
> (UNWTO, 2016, p. 10)

A new generation of travellers – Gen Z – has entered the youth travel market. Over a third of the world's population counts themselves as Gen Z, surpassing Millennials as the most populous generation on earth in the near future (Deloitte, 2020). For example, Gen Z makes up one-fifth of the population in North America and is the most racially and ethnically diverse generation in the country's history (World Economic Forum [WEF], 2018). Thus, Gen Z represents a fast-growing consumer cohort worldwide with far-reaching implications in the influence and spending power in tourism, recreation, and hospitality. The study by the European Travel

DOI: 10.4324/9781003289586-2

Commission (2020) on Gen Z travellers offers some interesting travel preferences and behaviours among this young cohort of travellers:

- European destinations are primarily chosen by this generation because of their safety and security;
- Their main purpose for travelling is vacation (70%);
- Gen Z are both Immersive Explorers, due to their interest in urban culture, gastronomy, and traditions, and City Life Enthusiasts, due to their interest in shopping and nightlife while on holiday;
- They are more likely to return to a destination over time.

There is an estimate that Gen Z accounts for $29 to $143 billion in direct spending and, as a result, they substantially influence purchases made by their families and households (Forbes, 2018). Gen Z is estimated to have a purchasing power five to six times greater than Generation X, which makes it a category of great interest to marketers (Dabija et al., 2019). Such purchasing power is also significant in tourism and travel and as the report of 2018 Expedia Group™ Media Solutions indicate that Gen Z travellers take 2.8 leisure trips per year, which is only just behind Millennials (three leisure trips per year) who have more income. While Millennials have previously been known as the driving power in the travel industry, Gen Z is assumed to be the mantle of becoming the 'new buzzword' in tourism. Family travel choices and decision-making are heavily influenced by Gen Z, and as more members of this generation join the workforce and have more disposable income, their preference for travel as well as their expanding budgets are expected to provide several marketing opportunities in the future (Expedia Group, 2018).

Gen Z is expected to play a significant role in revitalizing the travel and hospitality industry post-COVID-19 pandemic. According to a study by Contiki, 63% of Gen Zers and Millennials want to travel post-pandemic, while 58% would go immediately even if they had to pay for quarantine upon return (Cowling, 2021). According to Robinson and Schänzel (2019), Gen Z values social and environmental consciousness, mobile connectivity, and authentic local experiences when they travel. Other studies portray Gen Z as an environmentally conscious and sustainability-oriented generation (Dabija et al., 2019; Sharpley, 2021; Djafarova & Foots, 2022; Salinero et al., 2022; Prayag et al., 2022; Liu et al., 2022). Due to the fact that Gen Z will probably face the greatest environmental challenge in the future, it is widely regarded that this cohort will need to adopt more sustainable lifestyles and sustainable consumption patterns that may, of course, have profound implications for travel and tourism. Therefore, there is a need to create a clearer and more precise understanding of the mechanisms that underlie the consumer behaviour of Gen Z with respect to sustainable products, as well as for businesses to consider the entire consumer market.

Though Gen Z consumers have grown significantly in recent years and have become increasingly important to the global tourism industry, there has been relatively little empirical research on this generation in the tourism literature, which stands in marked contrast with other market segments (i.e., seniors' tourism) or Generation Y (Millennials). Hence, this introductory chapter contributes to the understanding of this significant tourism market and sheds light on their characteristics, behaviour, mobility, and sustainable consumption practices, as well as their overall role as emerging consumers and producers of new experiences of tourism. The chapter concludes by outlining the structure of the book.

Who and what is Gen Z?

Strauss and Howe (1997, p. 61) defined a generation as "an aggregate of all people born roughly over the span of a phase of life, who share a common location in history, and hence, a common collective persona". In a similar vein, Kupperschmidt (2000, p. 66) defined the concept of generation as an "identifiable group that shares birth years, age location, and significant life events at critical developmental stages". According to generational theory, the traits, values, attitudes, desires, and aspirations of members of each generation are distinct (Strauss & Howe, 1997). For Li et al. (2013, p. 148), a generational theory is "a dynamic socio-cultural theoretical framework that pinpoints patterns and propensities at an aggregate level across the generational groups, rather than for the individuals". While Gen Z is the demographic cohort that succeeds Millennials and precedes Generation Alpha, this is the one with the most difficulty defining its age range. Despite the lack of unanimity regarding the definition, Gen Z is usually regarded as comprising those born between the late 1990s and the late 2000s (Corey & Grace, 2019). A number of terms have been used to describe this first generation of the twenty-first century and these are summarized in Table 1.1.

Gen Z grew up in a world that is fundamentally different from that of previous generations. In addition to being large and heterogeneous groups (Corbisiero & Ruspini, 2018), they are the first generation born into a digital environment and

Table 1.1 Terms used to define Gen Z

Term	Reference
'I' Generation	Fyock et al., 2013; Horovitz, 2012
Net Generation	Corbisiero & Ruspini, 2018
Pivotals	Fromm & Read, 2018
The Uber Generation	Koulopoulos & Keldsen, 2016
Digital natives	Vojvodic, 2019
Net Generation	Horovitz, 2012
Hyper-connected generation	Haddouche & Salomone, 2018
Instant-gratification generation	Rota, 2017
Zoomers	Corey & Grace, 2019
Post-Millennials	Fry & Parker, 2018; Corey & Grace, 2019
Centennials	Capatides, 2015

are widely viewed as 'digital natives' since they have never lived without Internet access, making them a hyper-connected generation (Vojvodic, 2019). According to a report by McKinsey & Company (2018), Gen Z are digital natives who grew up with the Internet. They have distinct buying habits, where personalization and self-identity play a larger role than in previous generations. Haddouche and Salomone (2018, p. 69) describe the Gen Z cohort as a "new sociological category, nourished by the information technologies, the internet and the social networks". Wood (2013) characterizes Gen Z as consumers on four pillars: (a) an interest in new technologies, (b) an emphasis on user-friendliness, (c) a desire to feel safe, and (d) a desire to temporarily escape the realities they face. It is possible that all of these may be significant to varying degrees given that Gen Z has been exposed to many political, social, technological, environmental, health, and economic developments during their brief lifetime (Sakdiyakorn et al., 2021). As such, some argue that these worldwide events influenced Gen Z's sense of collective consciousness and contributed to the development of human values, such as universalism, altruism, self-direction, accomplishment, and security (Çalişkan, 2021). Although such events are experienced by all generations, technology is perhaps the most important characteristic of Gen Z. The influx of Gen Zers is therefore expected to lead to major changes in consumer travel, hospitality, and leisure demand. Understanding Gen Z travellers, their concerns, and their motivations is essential to assessing how such changes in tourism demand may affect global tourism, which will be discussed in the following sections.

Gen Z travel behaviours

As noted earlier, the Gen Z demographic represents nearly one-third of the global population which makes this market one of the fastest-growing travel segments in tourism. Haddouche and Salomone (2018) view Gen Z as sophisticated, hard to keep, and having high travel expectations. Key characteristics describing Gen Z tourists' behaviour were identified by Mignon (2003, cited in Bǎltescu, 2019): last-minute decisions, continuous search for opportunities, use of word-of-mouth recommendations to choose their destinations, and increasing use of low-cost services. Gen Z has been "an allegedly environmentally aware group" (Sharpley, 2021, p. 99) and supposedly more open-minded and culturally aware than previous generations, due to its persistent online presence as a digital native (Nikolić et al., 2022). Sharpley (2021) argues that:

> the younger, post-millennial generation in particular displays environmental credentials that point to a fundamental shift towards more sustainable consumption. Hence, the potential for reducing overall levels of tourism consumption might be manifested in the attitudes of post-millennial towards tourism consumption.
>
> (p. 101)

Gen Z is already changing the way the tourism and hospitality industries operate. Travel is a priority for Gen Z and a report by Booking.com (2019) revealed some interesting habits of Gen Z travellers: 79% of Gen Zers travelled abroad before the age of 15 and 65% chose 'travel and seeing the world' as the most significant way to spend their money (Booking.com, 2019). Clearly, this group of travellers is passionate about travelling and experiencing new places, although the results may also reflect something of Booking.com's own demographic. Gen Z travellers are also socially and ecologically responsible, mobile-first, and seeking authentic local experiences (Wee, 2019). As Haddouche and Salomone (2018, p. 70) note:

> Young people expect a great deal from their travels. The consumption of stays often translates into a hedonistic behaviour: tourism is at the same time a moment of conviviality, of socialisation, of implication and of empowerment. All these needs are reflected in specific purchasing behaviours: last-minute decisions, search for opportunities, use of word-of-mouth recommendation sources to choose their destinations, increasing use of low-cost service recommendation sources to choose their destinations, increasing use of low-cost services.

In their study, Robinson and Schänzel (2019) found that multiple factors determine Gen Z's travel experiences in a destination. They conclude that Gen Z's travel patterns and reasons are not fundamentally different from those of previous generations. However, the authors note that while the reasons and/or patterns may be similar, contemporary factors can shape the experiences of a generation, and traveller expectations and experiences have been altered by the advancements in technology (Internet, social media, smartphones). A similar conclusion was reached by Monaco (2018) regarding the technological differences between Italian Gen Y and Gen Z. According to this study, almost exclusively, Gen Z uses the web to make purchases and make reservations, share experiences via social networks, and use instant messaging and chat applications while using smartphones and tablets, which has led to their higher likelihood of leaving online reviews than other generations. In contrast, Gen Y makes greater use of personal computers, even though they are using mobile devices. With regard to Gen Z and the use of technology, Haddouche and Salomone (2018) found that:

> Although it is often presented as a narcissistic generation, seeking to put forward their 'selves', for example by posting selfies, this study reveals that Gen Z seems to show a great modesty during their tourist experiences.
>
>   (Haddouche & Salomone, 2018, p. 69)

The study of Tavares et al. (2018) on Gen Z travellers in Belo Horizonte, Brazil, indicated that while older generations prefer a relaxing vacation, younger generations seek out engaging experiential activities and tend to be environmentally conscious while still possessing a high level of open-mindedness.

What makes Gen Z tourists different from other travel segments?

As discussed earlier, Gen Zers have unique characteristics that set them apart from preceding generations. The values and beliefs of each generation are influenced by factors such as their surroundings, the current political, economic, and cultural context, as well as historical events. Each generation is characterized by predictable characteristics derived from events during their formative years (Benckendorff & Moscardo, 2013). According to Parment (2013), different events and interactions differentiate generational cohorts within a population, resulting in shifts in their beliefs, behaviours, and predispositions. Similarly, Jones et al. (2009) contend that generational differences are shaped by key historical events that occurred during a cohort's transition into adulthood. As Benckendorff et al. (2009) observed, the distinct and special pattern of beliefs, perceptions, and behaviours can have a significant and lasting impact on a generation's ability to respond to and create changes in the social and economic environment, including the tourism industry (Hall, 2005). Table 1.2 illustrates the context in which different generation cohorts emerged and compares the five generations of Traditionalists, Baby Boomers, Gen X, Millennials (Gen Y), and Generation Z regarding their attitudes, consumption patterns, and events that shaped their identities and behaviour.

From a generational perspective, an individual's behaviour is shaped by their experiences and their environment as they grow up. While Traditionalists grew up during World War I and II and experienced the Great Depression, Baby Boomers grew up during the Cold War era of rapid economic growth. Internet and digitalization shaped both Gen Z and Millennials. White (2017) argues that Gen Z's identity and life skills have been shaped by a socio-economic environment that is unstable, turbulent, and ambiguous. This echoes Robinson and Schänzel's observation (2019) that Gen Z grew up during a period marked by economic decline, inequality, job instability, and social media. Gen Zers, however, were born at a time when online shopping, smart devices, and social media were well-established and a 'natural' part of the home environment, whereas older Millennials (Gen Y) grew up before the Internet's rapid rise. Gen Z was also called a generation of digital natives by Stillman and Stillman (2017) who called it the first generation born into a digital world that corresponds to any physical element (people and places). Furthermore, a number of key issues that began in the early twenty-first century, such as globalization, global warming, the financial crisis, terrorism, the technology revolution and more recently the COVID-19 pandemic, have had a profound effect on Gen Zers' attitudes and beliefs. According to a recent survey (Simon, 2021), 83% of Gen Z Americans are concerned about the planet's health and say the quality of their environment affects their health and well-being. The same survey also highlights climate anxiety among the Gen Z cohort and reports that climate change is taken

Table 1.2 A comparative analysis of generations

Generation	Estimated % of global population 2017	Life-defining events	Behavioural traits	Consumption	Communication media	Attitude towards technology	Aspiration
Traditionalists (silent/maturists) ≈1900–1945	9%	• World War I and II • Great Depression • Fascism	• Loyalty • Hard work • Discipline • Value Authority • Conservative		Formal letter	Largely disengaged	Home ownership
Baby Boomers ≈1946–1964	24%	• Cold War • Authoritarianism • Women's Rights • Civil Rights Movements	• Responsibility • Idealistic • Collectivist • Work-centric	• Vinyl and movies teenagers	Telephone	Early IT adaptors	Job security
Generation X ≈1965–1980	20%	• Capitalism • Vietnam War • Mutually Assured Destruction • Anti-apartheid movement • Palestinian liberation movement	• Efficiency • Individualist • Materialistic • Competitive • Self-Reliance • Flexible	• Brands and cars • Vinyl and cassette • Luxury articles	Email and text message	Digital immigrants	Work-life balance
Generation Y/ Millennials ≈1981–1992	22%	• Globalization • Thatcherism and Reaganism • Technology and Internet Emergence • Terrorism • AIDS • End of Cold War	• Sociable • More Confident • Less Independent • Tech comfortable • Family-centric	• Festivals and travel • CDs • Flagships • Personal computers	Text message or social media	Digital natives	Freedom and flexibility
Generation Z ≈1995–2010	26%	• Post-Great Recession • Neoliberalism • Global warming • Digital Natives • Mobility • Social Networking	• Poor Communication Skills • Always connected • Multi-taskers • Realistic	• Unlimited • Ethical • Downloads • Mobile phones • iPod	Hand-held communication devices	Technoholics (entirely dependent on IT)	Security and stability

Source: After Francis & Hoefel, 2018; Neilson, 2010; Deloitte, 2020.

seriously by almost two-thirds of Gen Z compared to preceding generations. The top three environmental issues that concerned Gen Z Americans were air quality, water pollution, and plastic pollution (Simon, 2021). European Travel Commission's (2020) study of Gen Z travellers also noted that:

> while Millennials are often referred to as the 'Me-Generation' due to their individualist outlook on the world, many have observed that Gen Z'ers show greater interest in collectivist action. Whether through online forums or in the streets, Gen Z appears to have embraced activism and progressive ideas, showing a stronger interest than their older counterparts in themes, such as diversity, sustainability and personal empowerment.
>
> (p. 13)

Although interpretations do vary between studies, some general characteristics have been identified of the travel trends that characterize a generation. Table 1.3 offers an overview of travel trends across generations X, Y (Millennials), and Z. Although there are some similarities in vacation types there are differences in the amount of time taken up by travel while there are also clear differences in the 'flavour' of the travel that generations engage in. Also of significance is the use of different social media which suggests that Gen Z has a shorter period of engagement with specific messages even though the overall time used to engage in social media

Table 1.3 Travel trends across generations

Generation	Features	Travel days per year	Top vacation types	Most influential platforms
X	• Vacation-deprived road trip warriors • Travelling less frequently than other generations • Rely on word-of-mouth recommendations	26	• Relaxing • Visiting family • Sightseeing • Family play	• Facebook • Pinterest
Y	• Big spenders and travel the most • Prefer all-inclusive • Relaxing and romantic vacations	35	• Relaxing • Visiting family • Family play • Romantic getaway	• Facebook • Instagram
Z	• Budget-Conscious Open-minded • Bucket-list-oriented activities • Looking for off-the-beaten-track path locations	29	• Visiting family • Relaxing • Sightseeing • Special event	• Snapchat • Instagram • Facebook

Source: Expedia Media Solutions, 2018; Booking.com, 2019; Francis & Hoefel, 2018.

may be similar. Therefore, different generations utilise different platforms, a trend that is only likely to become more acute over time.

In their study on international visitors to the Canterbury region of New Zealand, Prayag et al. (2022) examined whether Gen Z is similar or different to three other generations (Gen X, Y, and Baby Boomers) in their environmental attitudes towards travel. The results of their study confirmed intergenerational differences in environmental attitudes and travel behaviours but also highlighted intra-generational differences. Gen Z tourists tended to be a heterogeneous group and were more likely to engage in sustainable practices related to resource-saving and buying local food compared to other generations. Furthermore, many of these assessments were conducted before COVID, so actual travel behaviour is also to be evaluated, although we recognize online usage has almost certainly been reinforced.

Gen Z and sustainability

Consumption decisions of many consumers are increasingly influenced by sustainability (Seyfi et al., 2022). In a highly globalized world, it is vital to understand how different generations consume and their level of awareness regarding sustainable consumption habits. Growing evidence suggests that Gen Z are aware of environmental and sustainability issues (Yamane & Kaneko, 2021; Seyfi et al., 2021; Sharpley, 2021; Prayag et al., 2022; Salinero et al., 2022; Seyfi et al., 2023). Sharpley (2021) contends that Gen Z is leading the way towards more sustainable tourism. Haddouche and Salomone (2018) demonstrated that Gen Z is sensitive to the concept of sustainable tourism, as shown by their environmental habits. Sharpley (2021) asserts that Gen Z serves as the spearhead of a consumer transition towards more sustainable tourism. Robinson and Schänzel (2019) also argued that environmental aspects of a destination are crucial to the travel experience of Gen Z. When it comes to travel, Gen Z is usually portrayed as being more concerned about the environment than previous generations. The study of Salinero et al. (2022) on pro-sustainable tourism behaviours of Gen Z travellers in the UK reported that three internal antecedents (awareness of consequences, ascription of responsibility and personal norms) and two external factors (social media engagement and membership of online community) have significant relationships with pro-sustainable tourism behaviours of this cohort of travellers. Whitmore (2019) argues that Gen Z's concern for the environment is reflected in their travel habits. They are interested in eco-friendly travel, sustainable hotel practices and contributing to the local community at their destination community. In a similar vein, Lin et al. (2022) focused on Gen Z's environmental pursuits and reported eudaimonic pursuits in shaping their sustainable tourism consumption. Casalegno et al (2022) also noted that Gen Z is more concerned about rising environmental issues than previous generations; a conclusion was also reported by Prayag et al. (2022) and Salinero et al. (2022). Budac (2014) suggests that Gen Z consumers are environmentally conscious and have an awareness of the environmental impact and carbon footprint of products. In their study on Gen Z Italian, D'Arco et al. (2023) found that Gen Z travellers are more likely to select sustainable transportation modes rather than eco-friendly hotels. Panzone et al. (2016) also reported that younger consumers are more concerned about the environment than older consumers, while older consumers are more inclined to take green consumer action. According to surveys reporting sustainable behaviour of the Gen Z cohort, Gen Zers have a strong sense of social responsibility when it comes to sustainable tourism and global environmental change. Among Gen Zers, 59% expressed genuine interest in leading change in sustainable development in the 2016 Masdar Gen Z global sustainability survey.

Although Gen Z holds strong environmental beliefs and attitudes, research shows that they are less engaged with actual environmental practices than prior generations (Giachino et al., 2022; Qiu et al., 2022). Bulut et al. (2017) reported in their study of Turkish consumers that Gen Z tends to consume unnecessary items. Hence, Hall (2013) emphasizes the importance of distinguishing between stated attitudes and actual behaviours when understanding sustainable consumption practices. Yet Haddouche and Salomone (2018) found that sustainable tourism is not a key concept for Gen Z based on their own tourist practices and use of social networks. The lack of clarity regarding the sustainability behaviours of Gen Z is indicative of the wider challenges businesses face in identifying the core factors that influence purchasing habits of younger generations. This was emphasized by Bulut et al. (2017, p. 599):

> consumption in 21st century is perceived by young generations as a means of communication and symbols of social status and from their perspective, the main purpose of consumption is to gain reputation by high exchange value.

Conclusion and outline of the book

As this chapter has outlined gaining an improved understanding of Gen Z and their behaviour is clearly significant for tourism research and practice. Given that modifying consumer behaviour, including those that arise from generational practices (Diprose et al., 2019; Bordian et al., 2022), is an integral element of sustainable consumption transitions, the collection of chapters in this book seeks to provide new insights into Gen Z attitudes, behaviours, and practices with respect to sustainability and tourism.

The book is divided into five sections. This chapter and the following chapter by Romera and Le Bigot contextualize the literature and some of the associated issues with respect to Gen Z and tourism. The second section consists of five chapters that examine Gen Z travel experiences, behaviours and patterns. Chapter 3 by Duedahl et al. looks at re-positioning Gen Z as drivers of sustainable development with a study of co-designing tourism with Gen Zs in Norway, while Suh (Chapter 4) compares the Gen Z environmental behaviour with that of other generations in Korea concluding that Gen Z is no less interested in practising a sustainable lifestyle and travelling to sustainable communities than older generations, although this does not mean Gen Z is more environmentally aware than older generations. In Chapter 5, Grénman et al. report on a study of Gen Z business students' environmental world view, environmental education, and environmental behaviour at a Finnish business school. The results show that tourism has become an essential part of Gen Zers' lives. Although the negative implications were recognized, giving up tourism entirely was not an option. Instead, maintaining sustainability in their daily practices was more reasonable than doing so in tourism, especially by seeking information, recycling, and considering their consumption choices. Chapters 6 and 7 provide a more specialized analysis of Gen Z and their tourism behaviour. In Chapter 6, Martins and da Costa compare the motivations and spatiotemporal behaviour of Gen Z and Gen Y backpackers in Porto in the north of Portugal. In Chapter 7, Monaco reports on Gen Z tourists' behaviour in Vesuvius National Park in Italy.

The third section examines Gen Z consumption behaviour in the hospitality sector. In Chapter 8, Saul and Heo report the results of their survey on Gen Z tourists' perception of hotels' green practices, while in Chapter 9, Vila-Vazquez et al. also report on a survey of Spanish students in relation to green practices and Gen Zers' behavioural intentions with respect to accommodation and lodging.

As noted in Chapter 1, there is a strong ethical dimension to Gen Z consumption. These issues are looked at in more detail in the fourth section which consists of three chapters. In Chapter 10, Orea-Giner focuses on issues of the Gen Z lifestyle in terms of food activism and sustainable travel behaviour. Chapter 11 details Gen Z tourists' perceptions of ethical consumption in Iran, while Chapter 12 provides a qualitative study of wildlife voluntourism and Gen Z.

The book concludes with a short chapter discussing issues and potential futures with respect to Gen Z and sustainable tourism consumption (Chapter 13).

Given the problems of environmental degradation, climate change, global heating, and biodiversity loss the development of sustainable tourism and the nature of consumption has become more important than ever (Hall, 2022). Consumers have a vital role to play in this by both their tourist decision-making and behaviour as well as their overall political consumerism when it comes to the acceptability of destination and business policies and practices. Critical to this is gaining a better understanding of generational dimensions of consumption given that generational practices with respect to sustainability and travel potentially last over the life course. Therefore, at this critical juncture for the sustainability of tourism it becomes vital to better understand the next and potentially largest generation of travellers the world has seen in the form of Gen Z. We, therefore, hope that this volume provides insights that will help tourism industry and researchers better understand the short- and long-term impacts of Gen Z and help develop more sustainable forms of tourism consumption.

References

Băltescu, C. A. (2019). Elements of tourism consumer behaviour of Generation Z. *Bulletin of the Transilvania University of Brasov. Series V: Economic Sciences*, *12*(1), 63–68.
Benckendorff, P., & Moscardo, G. (2013). Generational cohorts and ecotourism. In R. Ballantyne & J. Packer (Eds.), *International handbook on ecotourism* (pp. 135–154). Edward Elgar Publishing.

Benckendorff, P., Moscardo, G., & Pendergast, D. (Eds.). (2009). *Tourism and Generation Y*. CABI Publishing.

Booking.com. (2019). *From ambitious bucket lists to going it alone, Gen Z travellers can't wait to experience the world*. https://news.booking.com/from-ambitious-bucket-lists-to-going-it-alone-gen-z-travellers-cant-wait-to-experience-the-world/

Bordian, M., Gil-Saura, I., & Šerić, M. (2022). The impact of value co-creation in sustainable services: Understanding generational differences. *Journal of Services Marketing*. https://doi.org/10.1108/JSM-06-2021-0234

Budac, A. C. (2014). Strategic consideration on how brands should deal with Generation Z. *Journal of Communication Management, 66*, 6–14.

Bulut, Z. A., Kökalan Çımrin, F., & Doğan, O. (2017). Gender, generation and sustainable consumption: Exploring the behaviour of consumers from Izmir, Turkey. *International Journal of Consumer Studies, 41*(6), 597–604.

Çalişkan, C. (2021). Sustainable tourism: Gen Z?. *Journal of Multidisciplinary Academic Tourism, 6*(2), 107–115.

Capatides, C. (2015). Meet Generation Z. *CBS News*. www.cbsnews.com/pictures/meet-generation-z/

Casalegno, C., Candelo, E., & Santoro, G. (2022). Exploring the antecedents of green and sustainable purchase behaviour: A comparison among different generations. *Psychology & Marketing, 39*(5), 1007–1021.

Corbisiero, F., & Ruspini, E. (2018). Guest editorial. *Journal of Tourism Futures, 4*(1), 3–6.

Corey, S., & Grace, M. (2019). *Generation Z: A century in the making*. Routledge.

Cowling, C. (2021). *Voice of a generation: Our survey reveals your thoughts on the future of travel*. www.contiki.com/six-two/voice-of-a-generation-our-survey-reveals-your-thoughts-on-the-future-of-travel/

Dabija, D. C., Bejan, B. M., & Dinu, V. (2019). How sustainability oriented is Generation Z in retail? A literature review. *Transformations in Business & Economics, 18*(2).

D'Arco, M., Marino, V., & Resciniti, R. (2023). Exploring the pro-environmental behavioral intention of Generation Z in the tourism context: The role of injunctive social norms and personal norms. *Journal of Sustainable Tourism*. https://doi.org/10.1080/09669582.2023.2171049

Deloitte. (2020). *Welcome to Generation Z*. https://www2.deloitte.com/content/dam/Deloitte/us/Documents/consumer-business/welcome-to-gen-z.pdf

Diprose, K., Valentine, G., Vanderbeck, R. M., Liu, C., & McQuaid, K. (2019). Building common cause towards sustainable consumption: A cross-generational perspective. *Environment and Planning E: Nature and Space, 2*(2), 203–228.

Djafarova, E., & Foots, S. (2022). Exploring ethical consumption of Generation Z: Theory of planned behaviour. *Young Consumers*. https://doi.org/10.1108/YC-10-2021-1405

European Travel Commission. (2020). *Study on Generation Z travellers*. www.toposophy.com/files/1/ETC_REPORT_2020_vs8.pdf

Expedia Group. (2018). *A look ahead: How younger generations are shaping the future of travel*. https://info.advertising.expedia.com/hubfs/Content_Docs/Premium_Content/pdf/2018%20-%20Gen%20Z%20Travel%20Trends%20Study.pdf?hsCtaTracking=a63196b4-62b8-4673-93e2-7d3ad0dc73e3%7Cfd9915c8-dc7f-492d-b123–265614cef08a

Forbes. (2018). How much financial influence does Gen Z have? *Forbes*. www.forbes.com/sites/jefffromm/2018/01/10/what-you-need-to-know-about-the-financial-impact-of-gen-z-influence/#7ba6206a56fc

Francis, T., & Hoefel, F. (2018). *True Gen': Generation Z and its implications for companies.* https://www.mckinsey.com/industries/consumer-packaged-goods/our-insights/true-gen-generation-z-and-its-implications-for-companies

Fromm, J., & Read, A. (2018). *Marketing to Gen Z: The rules for reaching this vast – and very different – generation of influencers.* Amacom.

Fry, R., & Parker, K. (2018). *Early benchmarks show "post-millennials" on track to be most diverse, best-educated generation yet: A demographic portrait of today's 6-to 21-year-olds.* Pew Research Center.

Fyock, C., Finney, M. I., Robbins, S. P., & Thompson, L. (2013). *The truth about managing effectively.* FT Press.

Giachino, C., Bollani, L., Truant, E., & Bonadonna, A. (2022). Urban area and nature-based solution: Is this an attractive solution for Generation Z? *Land Use Policy, 112,* 105828.

Haddouche, H., & Salomone, C. (2018). Generation Z and the tourist experience: Tourist stories and use of social networks. *Journal of Tourism Futures, 4*(1), 69–79.

Hall, C. M. (2005). *Tourism: Rethinking the social science of mobility.* Pearson.

Hall, C. M. (2013). Framing behavioural approaches to understanding and governing sustainable tourism consumption: Beyond neoliberalism, "nudging" and "green growth"? *Journal of Sustainable Tourism, 21*(7), 1091–1109.

Hall, C. M. (2022). Tourism and the Capitalocene: From green growth to ecocide. *Tourism Planning & Development, 19*(1), 61–74.

Horovitz, B. (2012). After Gen X, millennials, what should next generation be. *USA Today.* https://abcnews.go.com/Business/gen-millennials-generation/story?id=16275187

Jones, I. R., Higgs, P., & Ekerdt, D. J. (2009). *Consumption and generational change: The rise of consumer lifestyles.* Transaction.

Kouloupoulos, T., & Keldsen, D. (2016). *Gen Z effect: The six forces shaping the future of business.* Routledge.

Kupperschmidt, B. R. (2000). Multigeneration employees: Strategies for effective management. *The Health Care Manager, 19*(1), 65–76.

Li, X., Li, X. R., & Hudson, S. (2013). The application of generational theory to tourism consumer behavior: An American perspective. *Tourism Management, 37,* 147–164.

Lin, Z., Wong, I. A., Wu, S., Lian, Q. L., & Lin, S. K. (2022). Environmentalists' citizenship behavior: Gen Zers' eudaimonic environmental goal attainment. *Journal of Sustainable Tourism.* https://doi.org/10.1080/09669582.2022.2108042

Liu, J., Wang, C., Zhang, T., & Qiao, H. (2022). Delineating the effects of social media marketing activities on Generation Z travel behaviors. *Journal of Travel Research.* https://doi.org/10.1177/00472875221106394

Masdar. (2016). *Engaging with the green generation: Masdar Gen Z global sustainability survey.* https://masdar.ae//media/corporate/downloads/wiser/masdar_gen_z_global_sustainability_survey_white_paper.pdf

McKinsey & Company. (2018). *"True Gen": Generation Z and its implications for companies.* www.mckinsey.com/industries/consumer-packaged-goods/our-insights/true-gen-generation-z-and-its-implications-for-companies

Mignon, J. M. (2003). Le tourisme des jeunes. Une valeur sûre. *Cahier Espaces, 77,* 16–25.

Monaco, S. (2018). Tourism and the new generations: Emerging trends and social implications in Italy. *Journal of Tourism Futures, 4*(1), 7–15.

Neilson, L. A. (2010). Boycott or buycott? Understanding political consumerism. *Journal of Consumer Behaviour, 9*(3), 214–227.

Nikolić, T. M., Paunović, I., Milovanović, M., Lozović, N., & Đurović, M. (2022). Examining Generation Z's attitudes, behavior and awareness regarding eco-products: A Bayesian approach to confirmatory factor analysis. *Sustainability, 14*(5), 2727.

Panzone, L., Hilton, D., Sale, L., & Cohen, D. (2016). Socio-demographics, implicit attitudes, explicit attitudes, and sustainable consumption in supermarket shopping. *Journal of Economic Psychology, 55*, 77–95.

Parment, A. (2013). Generation Y vs. baby boomers: Shopping behavior, buyer involvement and implications for retailing. *Journal of Retailing and Consumer Services, 20*(2), 189–199.

Prayag, G., Aquino, R. S., Hall, C. M., Chen, N. C., & Fieger, P. (2022). Is Gen Z really that different? Environmental attitudes, travel behaviours and sustainability practices of international tourists to Canterbury, New Zealand. *Journal of Sustainable Tourism.* https://doi.org/10.1080/09669582.2022.2131795

Qiu, H., Wang, X., Morrison, A. M., Kelly, C., & Wei, W. (2022). From ownership to responsibility: Extending the theory of planned behavior to predict tourist environmentally responsible behavioral intentions. *Journal of Sustainable Tourism.* https://doi.org/10.1080/09669582.2022.2116643

Robinson, V. M., & Schänzel, H. A. (2019). A tourism inflex: Generation Z travel experiences. *Journal of Tourism Futures, 5*(2), 127–141.

Rota, S. (2017). *The Gen Z answer key for business: The go-to guide for marketing to Generation Z.* Sky Rota.

Sakdiyakorn, M., Golubovskaya, M., & Solnet, D. (2021). Understanding Generation Z through collective consciousness: Impacts for hospitality work and employment. *International Journal of Hospitality Management, 94*, 102822.

Salinero, Y., Prayag, G., Gómez-Rico, M., & Molina-Collado, A. (2022). Generation Z and pro-sustainable tourism behaviors: Internal and external drivers. *Journal of Sustainable Tourism.* https://doi.org/10.1080/09669582.2022.2134400

Seyfi, S., Hall, C. M., Saarinen, J., & Vo-Thanh, T. (2021). Understanding drivers and barriers affecting tourists' engagement in digitally mediated pro-sustainability boycotts. *Journal of Sustainable Tourism.* https://doi.org/10.1080/09669582.2021.2013489

Seyfi, S., Hall, C. M., Vo-Thanh, T., & Zaman, M. (2022). How does digital media engagement influence sustainability-driven political consumerism among Gen Z tourists? *Journal of Sustainable Tourism.* https://doi.org/10.1080/09669582.2022.2112588

Seyfi, S., Sharifi-Tehrani, M., Hall, C. M., & Vo-Thanh, T. (2023). Exploring the drivers of Gen Z tourists' buycott behaviour: A lifestyle politics perspective. *Journal of Sustainable Tourism.* https://doi.org/10.1080/09669582.2023.2166517

Sharpley, R. (2021). On the need for sustainable tourism consumption. *Tourist Studies, 21*(1), 96–107.

Simon, S. (2021). *Gen Z is increasingly developing anxiety about climate change.* www.verywellhealth.com/gen-z-climate-change-anxiety-survey-5179490

Stillman, D., & Stillman, J. (2017). *Gen Z@ work: How the next generation is transforming the workplace.* HarperCollins.

Strauss, W., & Howe, N. (1997). *The fourth turning: An American prophecy.* Three Rivers Press.

Tavares, J. M., Sawant, M., & Ban, O. (2018). A study of the travel preferences of Generation Z located in Belo Horizonte (Minas Gerais – Brazil). *E-Review of Tourism Research, 15*(2–3).

UNWTO. (2016). *Global report on the power of youth travel.* www.wysetc.org/wp-content/uploads/2016/03/Global-Report_Power-of-Youth-Travel_2016.pdf

Vojvodic, K. (2019). Generation Z in brick-and-mortar stores: A review and research propositions. *Business Excellence*, *12*(3), 105–120.

Wee, D. (2019). Generation Z talking: Transformative experience in educational travel. *Journal of Tourism Futures*, *5*(2), 157–167.

White, J. E. (2017). *Meet generation Z: Understanding and reaching the new post-Christian world*. Baker Books.

Whitmore, J. (2019). How Generation Z is changing travel for older generations? *Forbes*. www.forbes.com/sites/geoffwhitmore/2019/09/13/how-generation-z-is-changing-travel-for-older-generations/?sh=4fe6eaf378f7

Wood, S. (2013). Generation Z as consumers: Trends and innovation. *Institute for Emerging Issues: NC State University*, *119*(9), 7767–7779.

World Economic Forum. (2018). *Generation Z will outnumber millennials this year*. www.weforum.org/agenda/2018/08/generation-z-will-outnumber-millennials-by-2019/

Yamane, T., & Kaneko, S. (2021). Is the younger generation a driving force toward achieving the sustainable development goals? Survey experiments. *Journal of Cleaner Production*, *292*, 125932.

Yeoman, I., Schanzel, H., & Smith, K. (2013). A sclerosis of demography: How ageing populations lead to the incremental decline of New Zealand tourism. *Journal of Vacation Marketing*, *19*(2), 91–103.

2 Sustainable tourism for sustainable development

Is Generation Z greener than Millennials?

Francisco Romera and Eugénie Le Bigot

Introduction

Generation Z (Gen Z) are the successors of the previous cohort called Generation Y (Gen Y) – or Millennials. In recent years, this generation has started to attract the attention of both practitioners and academics (e.g., Kirchmayer & Fratričová, 2018; Wiastuti et al., 2020). This generation of individuals born between the mid-1990s, 1996, and 2010 (Cho et al., 2018; Haddouche & Salomone, 2018) is now reaching legal adult age and thus becomes the latest active generation in terms of employment, production, and consumption. These factors, together with the fact that the members of Gen Z were born in a highly globalised and connected world and are considered technological natives, give them unique characteristics that differentiate them from their predecessor generations, such as Millennials (Robinson & Schänzel, 2019) who grew up with new technologies and with the expansion of market globalization. As for their way of perceiving social and environmental sustainability, the literature has highlighted the ability of Gen Z to empathise with environmental and social issues (Dabija et al., 2019; Seemiller & Grace, 2016). This may be due to the fact that they have grown up involved in a specific socio-political era (Seemiller & Grace, 2019). In addition, new generations have from a very early age witnessed two major global sustainable development agreements, such as the millennium development goals (MDGs) in 2000 and the Sustainable Development Goals (SDGs) in 2015. This generation also plays a significant part in tourism and travel (European Travel Commission, 2020). However, the academic literature on tourism is now starting to study Gen Z (Robinson & Schänzel, 2019). Some more research should be done to explore their relationship to sustainable development. In turn, this paper provides new insights into Gen Z's perception of sustainable tourism. Nevertheless, there exist some differences between Millennials and Gen Z (Kaplan, 2020). Millennials are characterised by an accumulated, materialistic, and consumer culture that primarily results from technological innovation (Hanzaee & Aghasibeig, 2010). Even recent works have pointed out that Gen Z individuals have a totally different behaviour in terms of sustainable orientation since they are behaving in a greener way than previous cohorts such as Generation X or Millennials (Dabija et al., 2019; Kaplan, 2020).

DOI: 10.4324/9781003289586-3

Tourism is a social phenomenon (Theobald, 2016; Yudina et al., 2016) that can generate employment (Sivesan, 2020). Sustainable tourism generates employment in the regions where it is carried out, as well as greater cultural development and awareness of the importance to conserve the natural heritage. It can also contribute to sustainable development. The new generations will be the guardians of sustainable tourism since many tourism-related activities worldwide are intended for them. Thus, Monaco (2018) asserts that they will be the most active group in the tourism market throughout this decade. Although the events described earlier lead this generation towards more sustainable patterns, these behavioural traits may differ depending on the background of each individual. This research focuses on and analyses the relationship between the different backgrounds of Gen Z and Gen Y members as well as their commitment to more sustainable tourism models. To that end, a sample of 266 members of Gen Z and Y with different backgrounds from the Valencian Community, Spain, participated in this research, which is based on a quantitative methodology, namely on frequency and correlation analyses.

The main conclusions of this article demonstrate that although Gen Z has distinctive characteristics that bring it closer to a more sustainable form of tourism, the different profiles, and personal factors such as education or gender are determining factors in the involvement with sustainability. However, these results do not strongly evidence the difference between the various generations when it comes to sustainable tourism.

The subsequent work is organised as follows. The next section explores the literature review where the two generations are considered in context as well as the relationship between sustainable development and sustainable tourism. The research methodology will then be presented, leading to the analysis and discussion of our results. The main conclusions, along with recommendations and future lines of research, are finally developed.

Literature review

From Millennials to Gen Z

Different generations have different characteristics (Grasso, 2014). The world has changed over the last 30 years: globalisation, new innovations and technologies, and means of transport and communication have drastically evolved. The era in which we were born and the different events we go through in life shape us (Campbell et al., 2015) impacting our behaviours and perspectives (Magano et al., 2020). Gen Z includes those who were born between 1996 and 2010 (Seemiller & Grace, 2019; White, 2017). This is the first generation that came to life in this new globally connected world (Cilliers, 2017) and represents 32% of the global population which has a growing influence on the tourism and hospitality sectors (Miller & Wei, 2018).

Even though both Millennials and Gen Z have grown up in an environment of mass access to the Internet, together with the explosion of multinationals operating on the Internet, such as Amazon, eBay, Google, and Netflix in a more global and connected world in which new technologies are omnipresent, there are two important differences in the space and time of these generations. On the one hand, Millennials – born between 1981 and 1995 (Kim, 2018) – have grown with these changes since their childhood or adolescence in the case of the older members. They saw this new world rise and evolve and they had, at different levels, to adapt to it. On the other hand, Gen Z was born into this scenario, and its members did not have to adapt to it while growing up or to develop new skills since they are considered digital natives (Childers & Boatwright, 2021; Southgate, 2017). There are several indicators for this evolution. One striking example of it is the worldwide crossed border bandwidth that has grown from 4.7 Terabits per second (Tbps) in 2005 to 211.3 Tbps in 2014 (McKinsey & Company, 2016), meaning that the speed multiplied by almost 45 in less than ten years.

A second important difference that affects the way we perceive tourism practices is the visibility in day-to-day activities and education programmes of environmental problems in the different space–times where these generations have grown up. Millennials lived their youth within the framework of the MDGs (United Nations, 2000) while Gen Z was born immersed in this stage and has been influenced by more ambitious SDGs since 2015. Gen Z is the most well-educated generation ever (Parker & Igielnik, 2020). According to a report by the Pew Research Center (Fagan & Huang, 2019), in the specific case of Spain, over 80% of the people believe that the topics related to sustainability such as climate change are major threats. These factors influence their involvement in socio-environmental movements, for example, the 'youth for climate' movement.

The case of Spain, patterns of consumption

Spain is a country with an important weight in the European and world tourism sectors. Spain currently has a population of 47,432,805 people, as of January 2022. In turn, according to Spanish statistics, of the total number of people living in Spain, more than 7 million (M) belong to Gen Z, 8M are Millennials and more than 9M belong to Gen X. This distribution indicates that almost 30% of the total are members of the most recent active generations, Gen Z and Millennials.

The Spanish case is not extremely different from the rest of developed countries where globalisation and digitalisation are equally influential. However, some characteristics of the Spanish case indicate a change in consumption patterns. For example, only 3.7% of young people between 20 and 24 years old read newspapers on a daily basis. This percentage increases with age, being 9.5% among 25- and 34-year-olds and 15.7% among 35- and 44-year-olds or 19.8% among 45- and 54-year-olds (Statista, 2021).

Other characteristics of the Spanish Gen Z are their preferences when it comes to finding a job. According to Deusto Business School and ATREVIA (2016), for this generation, the most important thing is to find a good working environment above salary. In terms of brand consumption, this same report revealed the importance of price as the first evaluated characteristic when choosing a brand, followed by immediate response to needs and thirdly, responsibility towards the environment.

Finally, when they require information, the younger generation chooses the Internet as the main means of information on consumer products. Overall, the behaviour of Gen Z in Spain towards sustainable tourism is a continuation of the Millennials' generation.

Sustainable development and sustainable tourism

Sustainable development was defined as a "development that meets the needs of the present without compromising the ability of future generations to meet their own needs" (Brundtland, 1987, p. 43). This definition considers three key elements: economic growth; social inclusion; and environmental protection. At the same time, these three elements have generated a range of concepts such as the triple bottom line, or 3Ps, People, Planet, Profit (Elkington, 1994). As previously mentioned, from the beginning of this millennium, two important agreements dealing with sustainable development have been on the roadmaps of governments worldwide in The Sustainable Development Agenda. The agenda 2030 includes 17 goals and 169 targets to maintain sustainable development and tourism as an economic and social activity (UNWTO & UNDP, 2017). Thus, goal 8 'Decent Work and Economic Growth', target 8.9 states: "by 2030, policies should be developed and implemented to promote sustainable tourism that creates jobs and promotes local culture and products" (UNWTO & UNDP, 2017, p. 61).

Tourism is a global industry involving hundreds of millions of people (Mason, 2016) and is a source of richness for the receptor country. Travel and tourism contribution to gross domestic product (GDP) was 1,065 BN US Dollars contributing to employment with over 35 million jobs in Europe during 2021 (World Travel & Tourism Economic Council (WTTC), 2021) and over 30 million international passengers visiting Spain during the last year (La Moncloa, 2021). Unfortunately, even if positive impacts emerge, tourism and the expansion of tourist traffic give rise to overtourism. Overtourism can be understood as "the situation, in which the impact of tourism, at certain times and in certain locations, exceeds physical, ecological, social, economic, psychological, and/or political capacity thresholds" (Peeters et al., 2018) and can have negative impacts on the environment, local communities, and economies (Maingi, 2019).

For the aforementioned reasons, it becomes essential to look for alternative ways of sustainable tourism. Moreover, the concept of sustainable tourism has been defined by the UNWTO as tourism that takes full account of its current and future economic, social, and environmental impacts, addressing the needs of visitors, the industry, the environment, and host communities. Sustainable tourism is an excellent opportunity to develop an awareness of conservation (Diedrich, 2007), improve host-community life (Lee, 2013), facilitate regional development (Whitford, 2009), socio-economic benefits (Sivesan, 2020), and provide cultural exchange between locals and visitors.

Gen Z, like the previous generation of Millennials, is paying particular attention to issues of sustainability and especially climate change. In connection with the COVID-19 crisis, more and more Gen Zers are choosing or aspiring to work from home and live in smaller cities, thereby reducing their environmental impact (Deloitte, 2022). In line with the behaviour of the Millennial generation, Gen Zers are trying to improve their environmental impact on a daily basis by, for example, buying second-hand clothes or ethical brands or promoting the consumption of local products (Deloitte, 2022). The literature also tends to prove that it is particularly women of Gen Z who pay the most attention to sustainable development in their daily consumption and practices (Brough et al., 2016; Gazzola et al., 2020).

Methodology

This research is based on a quantitative approach to analyse the behavioural patterns of Millennials and Gen Z regarding sustainable tourism by establishing comparisons between both generations, which are considered the most progressive generations of the last 60 years (Kaplan, 2020). The entire sample was collected over a single period from February to March 2022, following a cross-sectional analysis. The cross-sectional study design is one of the most well-known analysis designs, gathering data and representing the phenomenon studied at only one point in time (Olsen & St George, 2004). Survey-based research was chosen. For data collection, an electronically distributed questionnaire was used. The questions regarding perception and commitment followed a five-point Likert scale in which responders specify their level of agreement to a statement typically in five points: (1) Strongly disagree; (2) Disagree; (3) Neither agree nor disagree; (4) Agree; (5) Strongly agree (Taherdoost, 2019). The participants had to meet two selection criteria: to belong to Gen Y (Millennials) or Gen Z cohorts and be of legal age, with 18 years being the minimum age to take part in the study. This second criterion excludes all members of Gen Z who were born between April 2004 and 2010. A pilot study was conducted with seven people to improve the understanding of the questions and set up a time limit for responding to the survey. After having conducted the pre-test, several questions were rewritten to improve the quality of the survey. The survey contained a total of 19 questions in three sections, covering issues ranging from metrics and demographics characterising the respondents in terms of generation, gender, education, or professional situation, to more specific questions on topics like sustainable consumption, accommodation choices, or awareness of sustainable consumption. The survey was written in Spanish. At the same time, ethics were considered, and all the respondents were informed about the theme of the research, the 'voluntary' basis of the survey, and the possibility to withdraw at any point in the process. Moreover, the questions that could compromise the identity of the participants were not considered, thus ensuring the anonymity of the participants.

Sample

This research was carried out with individuals from the Valencian Community, Spain. The use of a single country to collect the sample helped this research in the following ways: First, it helped to avoid cultural and geographical biases since travel behaviour is influenced by the living context (Wilson et al., 2008). At the same time, these contexts involve historical, temporal, social or cultural aspects (Robinson & Schänzel, 2019). Therefore, this research considered two different generations, Gen Z and Millennials. The incorporation of Millennials in the sample will help to understand the possible differences. Second, Spain was chosen since it ranks first in the Travel and Tourism Competitiveness Index (World Economic Forum, 2019). A total of 268 responses were obtained, of which 266 were valid responses, 107 for Millennials and 159 for Gen Z. Questionnaires containing empty key answers were eliminated from the sample. Regarding the final sample, in absolute terms, there is an overrepresentation of women with 68% of the respondents being women and the remaining 32% being men.

Results and discussion

Some interesting characteristics of the sample are displayed in Table 2.1. While students from Gen Z represent 47.2% of this cohort, it drops to 5.6% in Millennials. However, most members of the Millennials category are part-time workers (66.4% of their cohort). This data is aligned with the current economic and labour situation of those under 35 in Spain (Civismo.org, 2021). Regarding the level of studies, both cohorts have similar percentages in terms of vocational training and undergraduate studies. They differ for primary and secondary school levels – where Millennials are more represented – and postgraduate levels where Gen Z is more represented.

By observing the gendered frequencies inside and between each generation, interesting results are found in the levels of agreement and disagreement with the questions asked, allowing us to achieve a higher level of understanding of the

Table 2.1 Respondents' demographics

		1995–2003 (Gen Z)		1981–1995 (Millennials)	
		%	*Frequency*	*%*	*Frequency*
Gender	Male	37.1	59	24.3	26
	Female	62.9	100	75.7	81
	Total	100%	159	100%	107
Professional situation	Student	47.2	75	5.6	6
	Studying and working	23.3	37	11.2	12
	Working Part time	23.9	38	66.4	71
	Working Full time	5.7	9	16.8	18
	Total	100%	159	100%	107
Level of studies	Primary and secondary education	22.6	36	26.2	28
	Vocational training	19.5	31	20.6	22
	Undergraduate	31.4	50	30.8	33
	Postgraduate	27.0	43	22.4	24
	Total	100%	159	100%	107

behaviour patterns. For each question, the level of agreement is considered high when the participants answered 4 or 5 and low when they answered 1 or 2. They were grouped into a set of questions considering the degree of commitment and the perception of both cohorts regarding tourism. In this way, the first set of questions focused on consumption habits and preferences between global companies and local enterprises when making purchases at the destination. In the first question concerning consumption in large fashion stores and well-known restaurants, Millennials tended to have a higher rate of disagreement than Gen Z with 37% compared to only 23.9%. However, if the gender factor is added, this difference is more pronounced for women than it is for men. If the question is about daily products and food, both generations prefer small local stores to large and well-known stores. This is in line with previous research indicating that Gen Z often seeks authentic local experiences (Monaco, 2018; Wee, 2019). However, this sample shows that Millennials are slightly more likely (57%) to consider local stores than Gen Z (48.4%). This result could be explained by economic factors. Indeed, Gen Z individuals included in this sample are younger and often still students: 70.5% in total against 16.8% for Millennials, making them more precarious. The gender factor is again of interest in the matter of daily products and food consumption patterns. Indeed, for both generations, the percentage of women who prefer buying from local shops – in a tourism context – is much higher (58% for Gen Z, 63.8% for Millennials) than that of men (47% for Gen Z and 50.9% for Millennials). Given the importance of the economic factor, some previous studies have shown that younger generations are ready to pay more to experience more sustainable tourism models (Dabija et al., 2019). However, in this research, the respondents from both cohorts indicated that the price was a determining factor in planning their vacations and accommodation. It should be noted that the question about consumption in global brands includes two global clothing brands and two global food-restaurant franchises as an example of the type of establishments we were asking about. The question regarding daily products and food brand stores follows the same criteria.

Another interesting result is the answers to the next set of questions on perceptions of tourism. The questions followed the same structure but considered three different impacts: *Tourism has an economic impact on the destination*; *Tourism has a social impact on the destination*; and *Tourism has an environmental impact on the destination*. Both generations strongly agreed on the economic impact with over 80% of the respondents. However, this level of agreement is lower for the social aspects and even lower for the environmental aspects. This may suggest that there is a need for policies and educational schemes to boost the awareness of the social and especially of the environmental repercussions of tourism in the choice of a destination. In terms of gender, men are less aware of the environmental impacts in both generations with around 92% of agreement, agree or strongly agree, for women of both Gen Z and Millennials against 81.4% for Gen Z men and 84.7% for Millennial men. However, there is no relevant difference between Gen Z and Millennials. Regarding the behaviour of these cohorts involving the management of their waste when in tourism, strong and very strong degrees of agreement were found when they were asked if they always put their waste in the bin or litter when doing tourism. This behaviour can be influenced by different social characteristics or prosocial behaviours (Dovidio et al., 2017). It should also be noted that Gen Z had a higher degree of agreement than Millennials.

Finally, regarding the correlation analysis, the results revealed that the level of education has a direct influence on the behaviour of individuals. Most often, one notices that students and participants with primary and secondary studies have a lower degree of awareness of the impacts of tourism and of the need to consider a more sustainable form of tourism. In turn, the level of studies and consumption in large companies, restaurants and fashion shops were found to be moderately positively correlated, $r(104)=.251$ in the case of the Millennials and $r(157)=.204$ in the case of Gen Z, $p < .001$. Therefore, when studying the issue further, one notices that there is less predisposition to consume in large food or clothing franchises while doing tourism. Small local establishments are preferred. Other significant correlations were observed considering the level of studies: the impact of tourism at the destination is moderately positively correlated with economics: 206, for Millennials and 203 for Gen Z, $p < .005$ and social .200 for Millennials, $p < .001$. Although a positive correlation with the use and deposit of waste in an appropriate receptacle was found among Millennials, these correlations could not be confirmed in the case of Gen Z. Even though previous studies indicate that Gen Z may adopt a greener lifestyle (Dabija et al., 2019; Valentine & Powers, 2013), this research shows that this is not as significant as desired and some variables, such as the level of studies, have a great influence on sustainable tourism.

Conclusions, contributions, limitations, and future lines of research

This study contributes to the discussion of whether the new cohorts play an important role in sustainable tourism. Millennials and Gen Z are considered to have been the most progressive generations of the last 60 years (Kaplan, 2020). However, each generation was shaped by different situational factors (Scholz, 2019). This chapter offers interesting results comparing the behaviour and commitment between two generations that represent the largest percentages of the population worldwide (Wood, 2018).

In the first place, this research revealed that gender is an important factor when tackling this issue. Women had a better understanding of and commitment towards sustainable development and sustainable tourism. This gender difference is more pronounced when it comes to the environmental pillar of sustainable development than when it comes to the economic and social pillars. This result is common to both cohorts.

Secondly, by analysing the level of study, this work reveals that university students and graduates have a greater awareness of their role as key actors in sustainable tourism development and of their contribution to sustainable development. It is important to promote and educate the current and future generations on sustainability (Bowser et al., 2014), to ensure the development of tourist destinations in the long term (Glover, 2009). These results contribute to research addressing the relationship between university education and the commitment to sustainable tourism (e.g., Rezapouraghdam et al., 2022; Țigu et al., 2010).

Thirdly, this generation is entering the labour market, and their purchasing power in terms of tourism is limited. This influences the way they organise their trips and can be a barrier to sustainable tourism. In this study, 81.1% of Gen Z and 68.2% of Millennials reported that price is one of the most important variables when planning their touristic outings. This can be one reason why this generation tends to prefer accommodations facilities that are perceived as being cheaper, such as Airbnb.

In conclusion, in this chapter, we used a sample from Spain to explore different behaviours of Gen Z and Millennials considering sustainable tourism and patterns of consumption. Digitalisation, globalisation, virtualisation, and other major social events have affected us over the last years and have contributed to the development of distinctive capacities in Gen Z, and the awareness of the environment has taken a leading role in political agendas in recent decades. This has influenced education for sustainable development. However, an effort is still required to raise awareness among all members of Gen Z in terms of sustainable tourism.

Like all research, this research has limitations. In the first place, although a specific geographical area was chosen in order to minimise cultural and geographic bias, it constitutes a limitation in terms of the generalisation of the results considering that the other geographic and cultural areas could not be studied. This opens a door to future research that should focus on other geographical areas. As a second limitation, this research considers the level of education, age, and gender to find out about the differences between Gen Z and Millennials. However, other socioeconomic variables such as purchasing power or type of profession have not been considered. Future research should consider these types of variables. Thirdly, all members of Gen Z born between April 2004 and 2010 were excluded from this research. Future research should therefore consider this population. This would be an interesting contribution to the understanding of this generation. Finally, the sample was collected during the COVID-19 pandemic. This circumstance offers an opportunity to see if COVID-19 has influenced or will influence the behaviour of individuals when they travel.

References

Bowser, G., Gretzel, U., Davis, E., & Brown, M. (2014). Educating the future of sustainability. *Sustainability*, *6*(2), 692–701.

Brough, A. R., Wilkie, J. E., Ma, J., Isaac, M. S., & Gal, D. (2016). Is eco-friendly unmanly? The green-feminine stereotype and its effect on sustainable consumption. *Journal of Consumer Research*, *43*(4), 567–582.

Brundtland, G. H. (1987). *Our common future: The report of the world commission on environment and development*. Oxford University Press.

Campbell, W. K., Campbell, S. M., Siedor, L. E., & Twenge, J. M. (2015). Generational differences are real and useful. *Industrial and Organizational Psychology*, *8*(3), 324–331.

Childers, C., & Boatwright, B. (2021). Do digital natives recognize digital influence? Generational differences and understanding of social media influencers. *Journal of Current Issues & Research in Advertising*, *42*(4), 425–442.

Cho, M., Bonn, M. A., & Han, S. J. (2018). Generation Z's sustainable volunteering: Motivations, attitudes and job performance. *Sustainability*, *10*(5), 1400.

Cilliers, E. J. (2017). The challenge of teaching generation Z. *PEOPLE: International Journal of Social Sciences*, *3*(1), 188–198.

Civismo.org. (2021). *The wealth of young Spaniards has been reduced by 94% between 2005 and 2017, according to Fundación Civismo*. https://civismo.org/es/la-riqueza-de-los-jovenes-espanoles-se-ha-reducido-un-94-entre-2005-y-2017-segun-fundacion-civismo/

Dabija, D. C., Bejan, B. M., & Dinu, V. (2019). How sustainability oriented is Generation Z in retail? A literature review. *Transformations in Business & Economics*, *18*(2), 140–155.

Deloitte. (2022). *The Deloitte global 2022 Gen Z and millennial survey*. https://www2.deloitte.com/global/en/pages/about-deloitte/articles/genzmillennialsurvey.html

Deusto Business School & ATREVIA. (2016). *Generación Z: El último salto generacional* [Generation Z report: The last generational leap]. Deusto Business School & ATREVIA.

Diedrich, A. (2007). The impacts of tourism on coral reef conservation awareness and support in coastal communities in Belize. *Coral Reefs*, *26*(4), 985–996.

Dovidio, J. F., Piliavin, J. A., Schroeder, D. A., & Penner, L. A. (Eds.). (2017). *The social psychology of prosocial behavior*. Psychology Press.

Elkington, J. (1994). Towards the sustainable corporation: Win-win-win business strategies for sustainable development. *California Management Review*, *36*(2), 90–100.

European Travel Commission. (2020). *Study on generation Z travellers*. https://etc-corporate.org/uploads/2020/07/2020_ETC-Study-Generation-Z-Travellers.pdf

Fagan, M., & Huang, C. (2019). *How people worldwide view climate change*. Pew Research Center. www.pewresearch.org/fact-tank/2019/04/18/a-look-at-how-people-around-the-world-view-climate-change/

Gazzola, P., Pavione, E., Pezzetti, R., & Grechi, D. (2020). Trends in the fashion industry. The perception of sustainability and circular economy: A gender/generation quantitative approach. *Sustainability*, *12*(7), 2809.

Glover, P. (2009). Generation Y's future tourism demand: Some opportunities and challenges. In P. Benckendorf, G. Moscard, & D. Pendergast (Eds.), *Tourism and Generation Y* (pp. 155–163). CABI Publishing.

Grasso, M. T. (2014). Age, period and cohort analysis in a comparative context: Political generations and political participation repertoires in Western Europe. *Electoral Studies*, *33*, 63–76.

Haddouche, H., & Salomone, C. (2018). Generation Z and the tourist experience: Tourist stories and use of social networks. *Journal of Tourism Futures*, *4*(1), 69–79.

Hanzaee, K. H., & Aghasibeig, S. (2010). Iranian Generation Y female market segmentation. *Journal of Islamic Marketing*, *1*(2), 165–176.

Kaplan, E. B. (2020). The millennial/Gen Z leftists are emerging: Are sociologists ready for them? *Sociological Perspectives*, *63*(3), 408–427.

Kim, S. (2018). Managing millennials' personal use of technology at work. *Business Horizons*, *61*(2), 261–270.

Kirchmayer, Z., & Fratričová, J. (2018, April 25–26). *What motivates Generation Z at work? Insights into motivation drivers of business students in Slovakia* [Published version]. Innovation Management and Education Excellence through Vision 2020. Proceedings of the 31st International Business Information Management Association Conference. http://ibima.org/conference/31st-ibima-conference/

La Moncloa. (2021). *Spain reaches 30.2 million international air passengers up to November*. www.lamoncloa.gob.es/lang/en/gobierno/news/Paginas/2021/20211220_air-passengers.aspx#:~:text=La Moncloa.,to November %5BGovernment%2FNews%5D

Lee, T. H. (2013). Influence analysis of community resident support for sustainable tourism development. *Tourism Management, 34*, 37–46.

Magano, J., Silva, C., Figueiredo, C., Vitória, A., Nogueira, T., & Pimenta Dinis, M. A. (2020). Generation Z: Fitting project management soft skills competencies – a mixed-method approach. *Education Sciences, 10*(7), 187.

Maingi, S. W. (2019). Sustainable tourism certification, local governance and management in dealing with overtourism in East Africa. *Worldwide Hospitality and Tourism Themes, 11*(5), 532–551.

Mason, P. (2016). *Tourism impacts, planning and management.* Routledge.

McKinsey & Company. (2016). *Digital globalization: The new era of global flows.* www.mckinsey.com/~/media/mckinsey/business functions/mckinsey digital/ourinsights/digi talglobalizationtheneweraofglobalflows/mgi-digital-globalization-full-report.ashx

Miller, L. J., & Wei, L. (2018). Gen Z to outnumber millennials within a year. *Bloomberg.* www.bloomberg.com/news/articles/2018-08-20/gen-z-to-outnumber-millennials-within-a-year-demographic-trends

Monaco, S. (2018). Tourism and the new generations: Emerging trends and social implications in Italy. *Journal of Tourism Futures, 4*(1), 7–15.

Olsen, C., & St George, D. M. M. (2004). *Cross-sectional study design and data analysis.* College Entrance Examination Board. www.yes-competition.org/media.collegeboard.com/digitalServices/pdf/yes/4297_MODULE_05.pdf

Parker, K., & Igielnik, R. (2020). *What we know about Gen Z so far.* Pew Research Center. www.pewresearch.org/social-trends/2020/05/14/on-the-cusp-of-adulthood-and-facing-an-uncertain-future-what-we-know-about-gen-z-so-far-2/

Peeters, P., Gössling, S., Klijs, J., Milano, C., Novelli, M., Dijkmans, C., Eijgelaar, E., Hartman, S., Heslinga, J., Isaac, R., & Mitas, O. (2018). *Overtourism: Impact and possible policy responses.* European Parliament, TRAN Committee.

Rezapouraghdam, H., Alipour, H., Kilic, H., & Akhshik, A. (2022). Education for sustainable tourism development: An exploratory study of key learning factors. *Worldwide Hospitality and Tourism Themes, 14*(4), 384–392.

Robinson, V. M., & Schänzel, H. A. (2019). A tourism inflex: Generation Z travel experiences. *Journal of Tourism Futures, 5*(2), 127–141.

Scholz, C. (2019). The Generations Z in Europe: An introduction. In C. Scholz & A. Rennig (Eds.), *Generations Z in Europe* (pp. 3–31). Emerald Publishing Limited.

Seemiller, C., & Grace, M. (Eds.). (2016). *Generation Z goes to college.* John Wiley & Sons.

Seemiller, C., & Grace, M. (Eds.). (2019). *Generation Z: A century in the making.* Routledge.

Sivesan, S. (2020). Sustainable tourism development in Jaffna district. *Journal of Tourism & Hospitality, 9*(3), 1–5.

Southgate, D. (2017). The emergence of Generation Z and its impact in advertising: Long-term implications for media planning and creative development. *Journal of Advertising Research, 57*(2), 227–235.

Statista. (2021). *Periódicos: Distribución de lectores por edad en España en 2021.* https://es.statista.com/estadisticas/476755/distribucion-de-lectores-de-periodicos-en-espana-por-edad/

Taherdoost, H. (2019). What is the best response scale for survey and questionnaire design; Review of different lengths of rating scale/attitude scale/Likert scale. *International Journal of Academic Research in Management, 8*(1), 1–10.

Theobald, W. F. (2016). *Global tourism.* Routledge.

Ţigu, G., Andreeva, M., & Nica, A.-M. (2010). Education and training needs in the field of visitors receiving structures and tourism services in the lower Danube region. *Amfiteatru Economic Journal, 12*(4), 735–760.

United Nations. (2000). *Millennium summit of the United Nations.* https://www.un.org/en/development/devagenda/millennium.shtml

Valentine, D. B., & Powers, T. L. (2013). Generation Y values and lifestyle segments. *Journal of Consumer Marketing, 30*(7), 597–606.

Wee, D. (2019). Generation Z talking: Transformative experience in educational travel. *Journal of Tourism Futures, 5*(2), 157–167.

White, J. E. (2017). *Meet Generation Z: Understanding and reaching the new post-Christian world.* Baker Books.

Whitford, M. (2009). A framework for the development of event public policy: Facilitating regional development. *Tourism Management, 30*(5), 674–682.

Wiastuti, R. D., Lestari, N. S., Ngatemin, B. M., & Masatip, A. (2020). The Generation Z characteristics and hotel choices. *African Journal of Hospitality, Tourism and Leisure, 9*(1), 1–14.

Wilson, J., Fisher, D., & Moore, K. (2008). "Van tour" and "doing a contiki": Grand "backpacker" tours of Europe. In K. Hannam & I. Ateljevic (Eds.), *Backpacker tourism: Concepts and profiles* (pp. 113–127). Channel View Publications.

Wood, J. (2018). *Generation Z will outnumber millennials by 2019.* World Economic Forum.

World Economic Forum. (2019). *Travel and tourism competitiveness report 2019: Travel and tourism at a tipping point.* https://www3.weforum.org/docs/WEF_TTCR_2019.pdf

World Tourism Organization (UNWTO) and United Nations Development Programme (UNDP). (2017). *Tourism and the sustainable development goals – journey to 2030, highlights.* www.e-unwto.org/doi/pdf/10.18111/9789284419401

World Travel & Tourism Council (WTTC). (2021). *Travel & tourism economic impact 2021.* WTTC.

Yudina, E. V., Uhina, T. V., Bushueva, I. V., & Pirozhenko, N. T. (2016). Tourism in a globalizing world. *International Journal of Environmental and Science Education, 11*(17), 10599–10608.

Part II

Gen Z travel experiences, behaviours, and patterns

3 Re-positioning Generation Z as drivers of sustainable development

Co-designing tourism with local Gen Zs

Eva Duedahl, Hogne Øian, Monica Adele Breiby, Birgitta Ericsson, and Merethe Lerfald

Introduction

In this chapter, the gap between the rhetoric used about the need to involve Generation Z (Gen Z) and their actual involvement in operationalising the United Nations (UN) Sustainable Development Goals (SDGs) is addressed. As drivers of sustainable change (e.g., ethical consumption), Gen Z may represent a renewed opportunity for the participation of this generation in further involvement in sustainable tourism development. Accordingly, the main focus is alternative ways of enabling both the participation of and the collaboration with Gen Z to identify opportunities for operationalising SDGs in the tourism sector. Based on a Norwegian national park context, the analysis draws on a yearlong research process that involves local Gen Zs (aged 15–18 years) in co-designing sustainable solutions in a tourism development context.

An overarching question posed in this chapter is in what ways Gen Z may contribute to sustainability transitions even beyond consumption and work, thus exploring how the uniqueness characterising this generation interacts with the wider fabrics of contemporary society. In this case, the setting is sustainable tourism development at the local level. More specifically, a key challenge for tourism research and practice is how to facilitate dynamic spaces of 'doing together' to improve understanding of the relationships between Gen Z and tourism providers, local communities, and tourists (Haddouche & Salomone, 2018; Wee, 2019).

Today's youth, labelled Gen Z, are recognised as a new generation with distinct attitudes, values, and behaviours (Scholz, 2019; Seemiller & Grace, 2019). From a lens of generational theory (Howe & Strauss, 2000), a growing number of studies suggest how the changed lifestyles, attitudes and values of Gen Z in different ways can drive sustainability transitions (e.g., Goh & Okumus, 2020; Su et al., 2019; Yamane & Kaneko, 2021). Representing a generation characterised by unique approaches to consumption in terms of sustainability concerns (Djafarova & Foots, 2022), Gen Z is also increasingly involved in the supply side of the commodity market (Dabija et al., 2020). Furthermore, Gen Z is held to be at the forefront of communications and information technologies, social networking, and consumption (Haddouche & Salomone, 2018; Skinner et al., 2018). In addition, implications

DOI: 10.4324/9781003289586-5

of the features of Gen Z for education (Cilliers, 2017) as well as work relations (Gaidhani et al., 2019) have been explored.

What is important to explore further are the ways in which Gen Z could bring these changed attitudes into play to enable a more sustainable development. Yet, research has until now mainly been concerned with identifying and measuring the sector-specific changes that characterise Gen Z. Accordingly, the question of how to involve and engage Gen Z to enhance and harness their unique capacities, values and attitudes for sustainable developments is paid closer attention to in this chapter.

To approach this question, the analysis draws upon recent research on design practice and collaborative tourism design (e.g., Duedahl, 2021; Heape & Liburd, 2018; Jamal et al., 2021; Liburd et al., 2020; Rogal & Sànchez, 2018). It extends the current understanding by considering a process of co-designing tourism as a potential evoking of spaces for microstructures. A microstructure is a small-scale, heterogeneous network of people who join forces to realise a local ambition (Basten, 2011).

The research is based on a yearlong purposeful involvement of high school students in the municipality of Dovre in the south-eastern part of Norway. Dovre is a point of convergence for recreational and protected areas including parts of three national parks. Due to the government's recent introduction of a new policy for national parks of greater inclusion of tourism activities, the area of the national parks as well as an extension of the protected areas have become a contested landscape (Breiby et al., 2022; Hovik & Hongslo, 2017). In part, the controversies relate to moral assessments on which kinds of activities, actors, and materialities should be parts of the landscape (Flemsæter et al., 2018). This has the propensity of making the questions of sustainability both value-laden and complex as a wide range of stakeholder interests are involved in the management of both tourist development and protection.

The presentation of the case of co-designing is followed by an analysis, focusing on the emergence, manifestation, and wider social relations of Gen Z's microstructures, before discussing how Gen Z can be repositioned as drivers of sustainable tourism change. The chapter ends with conclusions and implications for future tourism research and practice. By analysing the emergence, manifestation, and wider social relations of Gen Z microstructures, the chapter contributes knowledge and capacity for tourism researchers and practitioners to meaningfully involve Gen Z in sustainable tourism development processes. It outlines how contextually relevant co-design processes, tools and interventions can be one way of enabling the participation of and collaboration with Gen Z – on their own terms – in imagining the greater purposes and contributions of tourism.

Against this backdrop, this chapter will also contribute to expanding the focus on Gen Z by exploring this generation's current and potential role as drivers of co-creational efforts of sustainable development within the sphere of tourism. The aim of this chapter is also to provide tourism researchers and practitioners with more knowledge and improved capacities to meaningfully involve and engage Gen Z in sustainable tourism development processes.

On the research project

This research was part of a larger multi-disciplinary research project entitled 'Sustainable Experiences in Tourism' (2018–2020) funded by the Competence, University, and Research Development Fund of the county of Oppland in Norway. Based on discussions with public and private national, regional, and local project partners, Dovre municipality became a study setting according to its geographical connection to multiple national parks, and its developmental and demographic challenges.

The research project is approved by the Norwegian Centre for Research Data (NSD) and complies with the EU's General Data Protection Regulation (GDPR, 2016/679/EU) adopted by the NSD in 2018. The research was not evaluated as sensitive, and additional Gen Z consent to participate was obtained beforehand from parents through teachers. To ensure GDPR compliance throughout the research, tape-recording was not an option. Thus, the analysis relied on notes and extracts from various forms of materials. To appreciate the complex co-evolution of problems and solutions when co-designing tourism, analysis was guided by abductive theorising (Jamal et al., 2021).

The study that forms the base of this chapter draws on recent research on design practice and collaborative tourism design (e.g., Duedahl, 2021; Heape & Liburd, 2018; Jamal et al., 2021; Liburd et al., 2020; Rogal & Sànchez, 2018). It extends the current understanding by considering a process of co-designing tourism as a potential evoking of spaces for microstructures. A microstructure is here understood as a "small-scale, heterogeneous network of people . . . who join their forces" to realise a local ambition "they really care about" (Basten, 2011, p. 136).

Co-designing ways of 'doing national park' with Gen Z

Tourism co-design is a co-generative and co-learning research and development endeavour (Liburd et al., 2017). By bringing into play a range of processes, methods, tools, and interventions, those involved may, with others, engage in a sustainable tourism development process that represents spaces of opportunities (Heape & Liburd, 2018). Co-designing tourism takes the point of departure in the different values, perspectives, resources, knowledges, experiences, capabilities, and world views of those involved (Duedahl, 2021; Rogal & Sànchez, 2018). Co-designing tourism with Gen Z is thus an attempt to both engender a process of participation and collaboration for sustainable tourism development. In this endeavour, Gen Z is enabled to identify opportunities for change which they importantly can identify with.

Microstructures do not initially exist or operate on scales too small to leverage noteworthy effects. In other words, they need to be facilitated, created, and made manifest (Basten, 2011). It is thus imperative to not only consider the emergence and manifestation of Gen Z microstructures but also consider their wider social relations. By working with possible variations in the ongoing social- and power relations, novelty and transformations may arise in the micro-detail of interactions between many people who collaborate and who (come to) care as latent opportunities for stewardship becoming with others (Liburd et al., 2020). Herein also lies the opportunity to enhance and harness the unique capacities, values, and potentialities of Gen Z in imagining the greater purposes and contributions of tourism.

This approach calls for careful preparation, in which strategies for meaningful collaboration and design research go hand in hand. A few distinctions are worth drawing before outlining how to co-design with high school students as representatives of Gen Z. Bratteteig et al. (2012) explain that methods are general guidelines for how to carry out a co-design process, whereas tools and techniques offer the more specific instruments or ways of doing and making, that can be engaged in a process of co-designing tourism. Tools, moreover, are tailored to provide interactive situations where different techniques can be used (Peters et al., 2020). Last, intervention is understood as referring to tourism co-design activities where the researchers intentionally take on the roles of interventionists, such as during workshops. The experimental co-design process took place both inside and outside the students' classrooms, with the aim of developing an innovative tourism co-design tool and a public exhibition. In the following, six iterative tasks taken to facilitate Gen Z micro-structures are highlighted.

First, the co-designing of tourism is proposed as an approach to facilitate the participation of and collaboration with Gen Z. Second, the unexplored concept 'to do national park' is introduced as a verb form into the co-design process with the aim of shifting the focus from considering what a national park 'is' according to a series of predefined criteria, to instead enabling Gen Z to re-envision what their national parks may alternatively 'become'. Third, the research and development process included an innovative tool for integrating the SDGs and tourism, whereby Gen Z was empowered to take on the role of tourism co-designers. Through their subsequent multimedia productions, the Gen Zers brought sustainable national park practices to the forefront, which were made public, discussed and circulated at an exhibition with politicians, tourism practitioners, and researchers. Fourth, to mobilise local Gen Z, two teachers responsible for a tourism elective course at a local high school were approached by outlining the aim of research with an invitation to step into potential microstructures. Hereby sustainable tourism development, including different ways of doing national parks, became integral parts of pupils' (aged 15–18) attuned curriculum. Upon discussion with the pupils, this step also entailed signing up for a future intervention designed in step two. Fifth, to involve and engage Gen Z students in co-designing tourism, an innovative tool integrating the SDGs and tourism was developed and coined as a 'tourism co-design puzzle' (photo 1). The term 'puzzle' as opposed to 'game' is used because it aligns better with how sustainable tourism development is understood in this chapter. That is, compared to a game the notion of a puzzle emphasises the relational and collaborative over the competitive, the processual over a set finishing line to be reached, and emergent over structured rules. As next elaborated, the puzzle comprises various forms of hexagons, which immediately encourage the creation of patterns between the involved Gen Z students (Figure 3.1).

Figure 3.1 The 'tourism co-design puzzle' tool

The co-design puzzle is about identifying latent opportunities for sustainable tourism development as those involved continually make, (re)make, and (de)make new traceable patterns of meaning through their changed or new ways of relating. The puzzle was subsequently tested and adjusted in the project group prior to engaging the puzzle to evoke microstructures with and among the Gen Z students.

The final iterative task consisted of a daylong workshop held with the aim of co-designing new ways of doing national parks for sustainable tourism development with local Gen Z using the tourism co-design puzzle. One researcher facilitated the overall workshop consisting of six groups of five Gen Z students. One researcher of the project group participated in each group. Initially, upon reflecting on what it meant to do a national park in Dovre, the Gen Z students picked what they individually evaluated as the five most central aspects and noted these (as words, symbols, drawings) on red hexagons. The Gen Z students then negotiated and identified emergent patterns of meaning by thematising their hexagons as a possible mapping of current tourism situations. Second, the Gen Z students introduced to each other an artefact (a meaningful memory from a national park) to highlight what they think is important when doing a national park. Further, they were asked to write down the five most central aspects of the green hexagons, which were then interweaved into the red hexagons generating new patterns of meaning to reveal possible tensions, conflicts and values related to current and desirable tourism futures. Third, the Gen Z students interweaved what they evaluated as the most relevant SDGs and sub-targets of actions into their puzzles. Lastly, based on the new insights, the Gen Z students crafted new ways of doing national parks (new initiatives, changes, actions, concepts) for sustainable development, which were then introduced and discussed in the plenum.

As part of the Gen Z student's curricular activities, the six microstructures continued advancing their identified, preliminary new ways of doing national parks for sustainable development. They went out and beyond, even using their leisure time, to deepen their learning and understanding of ways of doing national parks by, for example, interviewing each other, generating mind maps, and trying out their proposals. To further enable the participation of and collaboration with the Gen Z students in a way that better speaks to them, they were encouraged to create multimedia outlets and prepare to make them public in the fifth step of research.

The next idea, of doing a public exhibition, came in response to the question of how to better appreciate and introduce others to the Gen Z students' creative, visual, and digitalised co-design outcomes whilst generating new meaningful relations around their microstructures. A public exhibition, attended by politicians, tourism practitioners, researchers, and students, on what it is (and may become) to do national park from the perspective of Gen Z was subsequently held at the Inland Norway University of Applied Sciences. The lead of the research project held an inaugural speech before the six micro-structures in turn introduced and engaged the audience in their new ways of doing national parks for sustainable development.

Abduction views theoretical contributions as emerging from researchers entering the field with a theoretical base that is continuously attuned as research progresses through the identification of 'surprising' patterns from repeated empirical observations (e.g., Timmermans & Tavory, 2012). To search out such surprises from the emergence, manifestation, and wider social relations of Gen Z microstructures, various illustrative snippets of fieldwork were explored and discussed, such as quotes, co-design materials, and multimedia productions.

Findings and discussion

This section focuses on the emergence, manifestation, and wider social relations of Gen Z microstructures. Table 3.1 briefly outlines the six Gen Z microstructures, including their identified opportunities for new ways of doing national parks for the SDGs and sustainable tourism change.

By incorporating the SDGs into their current ways of doing national parks, the Gen Z students were afforded other lenses from which they could alternatively

Table 3.1 Overview of Gen Z microstructures and their new ways of doing national parks

Gen Z microstructures	New ways of doing national parks for sustainable tourism development
We lead the way	To enhance inclusive, sustainable development, the project "we lead the way" involves schools located in national parks. Each class in turn introduces others to challenges and opportunities to foster collaboration and knowledge sharing and take preventative measures towards current unresponsible tourism production and consumption.
Reduce the pressure	To approach current overtourism at certain sights, the initiative "reduce the pressure" gives way to innovating tourist inflows by channelling them to alternative sites whilst mitigating some of the adverse effects of trash, degradation of nature and loss of species habitats.
Dovre hike	Recognising human-induced impacts on nature, the aim of "Dovre hike" [an app] is to re-direct current tourism inflows focusing on the off-the-beat locations including a range of historic, cultural, and local features to protect nature and biodiversity.
Ferda memory	The app "Ferda memory" stirs current tourist flows, integrates climate change measures, and facilitates sustainable tourism production and consumption by enhancing tourists' competencies and knowledge about culture and nature, including rental of, for example, outdoor clothes, bikes, and electric cars.
A greener national park	To combat current inequalities related to the access to, and health benefits of, engaging with nature, the initiatives of "a greener national park" entails local lunchboxes and a green passport to enhance the competencies of tourists and visitors whilst protecting nature.
Explore Dovre	Seeking to protect nature, the concept of the "Dovre hike" educates local youth as tourist guides also with the responsibility to register current factors related to the monitoring of climate change and biological diversity.

explore and expand their current understandings in a quest for identifying future ways of doing national parks. These settings further afforded spaces for the consecutive emergence of microstructures. Aligning with the results of the new Future of Humanity survey (Amnesty International, 2019), most microstructures identified Stop Climate Change (SDG #13) and related Life on Land (SDG #15) and Sustainable Production and Consumption (SDG #12) as the most important matters to address when finding new ways of doing national park for sustainable development.

Gen Z students experienced climate change in different ways by referring to melting glaciers, natural disasters such as flooding, and the upward treeline expansion. Some students shared how these visible, fast, and major changes are both 'scary' and 'concerning'. Changes that intimately traced into Gen Z's concern for Life on Land (SGD #15) as conveyed through their priority to enhance biodiversity, sustain, and restore healthy ecosystems, including the endangered wild reindeer and its natural habitat in the mountain areas close to where they lived. Relatedly, the Gen Z students emphasised the unsustainability of tourism production and consumption in cases of overtourism at certain sights and attractions, in particular with reference to degrading nature.

By taking part in the co-design of tourism, the Gen Z students expanded the scope of tourism's contribution to sustainable development by incorporating other SDGs than those the tourism industry is traditionally confined to. A paradox emerges considering how the vast list of targets of actions and indicators are the exact means from which to measure tourism's 'successful' contribution to sustainable development. Some of the Gen Z students emphasised that by sharing their experience-based relationship to, and knowledge of the history and culture of the mountains, they could contribute to developing new and more sustainable ways of doing national parks.

By considering the process of co-designing tourism as a potential evoking of spaces for microstructures, this chapter highlights how microstructures emerged as Gen Z students brought into play their varying perspectives, experiences, capabilities and situated knowledges (Duedahl, 2021; Rogal & Sànchez, 2018) to address a series of local challenges that became intimately embedded into the global agenda of the SDGs. By doing so, a sense of current constraints and opportunities simultaneously arises (Jamal et al., 2021). This is also where the innovative potential of including Gen Z in sustainable tourism changes surfaces. They expressed that their involvement in a learning process about new ways of thinking about sustainable tourism development made them see the Dovre Mountain Area (thinking 'outside the box', as one student expressed it), with new eyes and recognise the latent potential of further actions.

> I feel that I have learned a lot about sustainable tourism development and about how to think 'outside the box'. I have also learned more about the Dovre Mountain Area and the latent potential that lies here waiting and how we might further work with it.
>
> (Local newspaper, Rudiløkken, 2019)

The students' teachers appreciated the opportunity for the local Gen Z to work with their immediate national parks in a way that they could identify with, had generated high levels of ownership, engagement, motivation, and creativity. In the local newspaper's report from the co-design event, one student stated the following:

> The national parks will always be part of our identity, even if we might not ourselves reside here in the future, we want everything to continue being here.
>
> (Local newspaper, Amundgård, 2019)

Other microstructures similarly found that they 'actually just want it to be as it is' with reference to preserving nature and wildlife, and in some cases even erase past generations' unsustainable impacts to 'get back what was here before'. In this endeavour, the Gen Z students oftentimes situated tourism within a humble nexus of anthropocentric awareness status-quo tourism development.

Considering a recent controversy over the use of electric bikes in the parks, some of the Gen Z students argued that it was awareness and behaviour that counted, not the equipment. Accordingly, bikers could be directed to more sustainable ways of biking in the national parks by offering phone applications informing about vulnerable flora and fauna, and favourable routes taking this into account. As novelty and transformation arose in Gen Z's microstructures, micro-detail of interactions (Liburd et al., 2020), new and shared values for 'reasonable', 'responsible', and 'sustainable' ways of doing national parks for sustainable development emerged. The various microstructures brought together preventative measures and notions of care by proposing slogans such as 'use your head and take care of nature', 'keep it clean', 'don't throw garbage; bring garbage' and 'leave nature in the same state as you would like to encounter it', 'enjoy nature without disturbing or degrading nature'. These slogans traversed into a steady concern to protect, preserve, enhance, and safeguard the culture, natural resources, biodiversity, and wildlife of the mountain area, bringing to light various notions of caring for their national parks. Hence, it appears as the sustainable actions of Gen Z are guided by their ability to transcend their own selfish interests in support of enhancing human and non-human others (Sakdiyakorn et al., 2021; Dabija et al., 2020). Adding to these values local Gen Z's views on care include notions of being 'other-regarding' or 'other-interested' (Jamal & Menzel, 2009), which both refer to the ability of humans to overcome self-interest in the care of others, one's community, and the non-human other, such as the natural world (Fennell & Cooper, 2020). Herein, also lies the opportunity to enhance the values of Gen Z in re-imagining in their own terms the greater role and functioning of tourism. It is from such positioning it may be possible to overcome common, quick-fix solutions that characterise much of the continued unsustainable production and consumption of tourism (e.g., Seyfi & Hall, 2021) and instead as encouraged by the United Nations (UN; 2016) to begin channelling the creativity and activism of Gen Z to a better world.

Local Gen Zs were not interested in attracting more 'out-of-village people' or 'city folks' as tourists who do not 'understand our nature' or who degrade as opposed to respect nature and who generate limited if any economic contributions to the area. As this is in line with the older generations' resentment of the transformation of local natural landscapes into a resource for the tourism industry (Øian, 2013; Skogen et al., 2017), it highlights the inherent limitation of much current research when *only* considering Gen Z as something distinct or unique, as opposed to considering Gen Z as part of the wider fabric of contemporary society and in relation to the various stakeholders in tourism (Dabija et al., 2020; Skinner et al., 2018; Törőcsik et al., 2014; Wee, 2019).

At the public exhibition, the Gen Z students presented and discussed their identified opportunities for ways of doing national parks for sustainable development in the Dovre Mountain Area with local politicians, tourism practitioners, researchers, and university students. At this event, they emphasised the importance of collaboration with older generations, local non-tourism stakeholders and the various stakeholders within the tourism industry.

Having entered a position that had empowered them as co-designers, these Gen Zs had come to perceive themselves as integral participants and collaborators in a sustainable tourism transition process in the Dovre Mountain Area as emphasised through their usages of 'us'. The Gen Z students highlighted how the 'way out' of the current unsustainable production and consumption of tourism will not be easy but possible if strengthening the means of operationalisation through collaboration based on 'greater levels of tolerance and adaptation' and 'openness' towards one another.

In part with the aim of serving seeds for future collaboration as a diffusion of microstructures, the Gen Z students introduced their own co-designed ways of doing national parks to others in several settings. First, they were invited to present their outcomes at an event arranged by the local municipality of Dovre and the national park destination management organisation. Second, they participated in the closing seminar of the overall research project. Third, the local high school plan to board a new class of Gen Zs on an 'innovation tourism lab train' passing through their Dovre Mountain Area.

The Gen Z students' contributions to more sustainable ways of doing national parks increased local awareness of sustainable tourism development in the protected mountain areas. Moreover, their involvement appeared to evoke a sense of prideful belonging among them that is often absent in public trajectories on the struggles and challenges faced especially in rural and mountain areas in Norway. Aided on its way also by the local newspaper and regional broadcasting radio reporting on the Gen Z students' new ways of doing national park for sustainable development, the process facilitated a shift in perception of their 'disengaged' to a positive perception of Gen Z as a corps of caring stewards advocating for the sustainable development of the Dovre Mountain Area.

Conclusion and implications for tourism research and practice

The momentum spurred by ongoing debates about Gen Z as drivers of sustainable change has created a renewed opportunity for this generation to participate in further involvement in sustainable tourism development. This research has explored an alternative way of enabling both participation of and collaboration with Gen Z to identify opportunities for operationalising the SDGs and sustainable tourism change. It has provided answers to questions of how to meaningfully enhance and harness the unique capacities, values, and potentials of Gen Z to leverage sustainable tourism change in a protected natural area.

The findings of this research implicate that a thoughtful re-positioning of Gen Z as drivers of sustainable tourism change is required to leverage change against the continued unsustainable consumption and production of tourism. Specifically, if the unique attitudes of Gen Z are to transform tourism (Robinson & Schänzel, 2019; Wee, 2019), research and practice are to shift away from an entrenched mindset of provisioning less unsustainable services or work practices to Gen Z as passive recipients towards an approach of empowering and enabling Gen Z to realise their potential with others. This distinction is essential if one is to address the widening gap between the values and principles of sustainable development and actual operationalisation in tourism practice (e.g., Sharpley, 2020). Only through the meaningful participation of and collaboration with Gen Z will it be possible for us to carry the 2030 Agenda for Sustainable Development forward. In view of recent publications advocating rethinking of boosterism paradigm that dominates the tourism industry (Hall, 2019; Breiby et al., 2021), in part inspired by the multiple motives of Gen Z tourists (Haddouche & Salomone, 2018), this chapter has sought to demonstrate the potentials of the involvement of Gen Z in sustainable tourism changes.

The analyses of the local Gen Z's participation in a co-design process demonstrate that generational theory can be a useful lens in sustainability research. However, its complexity must be recognised, and the study has the following limitations. First, the Scandinavian setting, relying on conceptualisations of sustainable development that are inherently Western, implies co-design practices that are inherently Eurocentric (e.g., Jamal et al., 2021). Second, despite this study's usage of the concept of Gen Z, the perspectives and experiences of Gen Z are highly heterogeneous both within and across countries (e.g., Seemiller & Grace, 2019). Third, the purpose of co-designing tourism is not to produce generalisable results to be applied to other tourism situations. Still, with some precautions, the tool and interventions of this study could be adjusted to and engaged with Gen Z in different locations and social contexts to enable their participation and collaboration whilst leveraging other variations of meaningful sustainable tourism change.

References

Amnesty International. (2019). *Climate change ranks highest as vital issue of our time – Generation Z survey.* www.amnesty.org/en/latest/news/2019/12/climate-change-ranks-highest-as-vital-issue-of-our-time/

Amundgård, F. (2019). Reiselivselevane lærar verdien av "å nasjonalparke". *Vigga Pluss.* www.vigga.no/nyheter/2019/09/19/Reiselivselevane-lærar-verdien-av-å-nasjonalparke-19980902.ece

Basten, F. (2011, January 13–15). *Microstructures as spaces for participatory innovation* [Published version]. Proceedings of the 1st Participatory Innovation Conference, University of Southern Denmark.

Bratteteig, T., Bødker, K., Dittrich, Y., Mogensen, P. H., & Simonsen, J. (2012). Methods: Organising principles and general guidelines for participatory design projects. In J. Simonsen & T. Robertson (Eds.), *Routledge international handbook of participatory design* (pp. 117–144). Routledge.

Breiby, M. A., Øian, H., & Aas, Ø. (2021). Good, bad and or ugly tourism? Sustainability discourses in nature-based tourism. In P. Fredman & J. V. Haukeland (Eds.), *Nordic perspectives on nature-based tourism: From place-based resources to value-added experiences* (pp. 130–142). Edward Elgar Publishing.

Breiby, M. A., Selvaag, S. K., Øian, H., Duedahl, E., & Lerfald, M. (2022). Managing sustainable development in recreational and protected areas: The Dovre case, Norway. *Journal of Outdoor Recreation and Tourism, 37,* 100461. https://doi.org/10.1016/j.jort.2021.100461

Cilliers, L. (2017). Wiki acceptance by university students to improve collaboration in higher education. *Innovations in Education and Teaching International*, *54*(5), 485–493.

Dabija, D. C., Bejan, B. M., & Puşcaş, C. (2020). A qualitative approach to the sustainable orientation of Generation Z in retail: The case of Romania. *Journal of Risk and Financial Management*, *13*(7), 152.

Djafarova, E., & Foots, S. (2022). Exploring ethical consumption of Generation Z: Theory of planned behaviour. *Young Consumers*, *23*(3), 413–431.

Duedahl, E. (2021). Co-designing emergent opportunities for sustainable development on the verges of inertia, sustaining tourism and re-imagining tourism. *Tourism Recreation Research*, *46*(4), 441–456.

Fennell, D. A., & Cooper, C. (Eds.). (2020). *Sustainable tourism: Principles, contexts and practices*. Channel View Publications.

Flemsæter, F., Gundersen, V., Rønningen, K., & Strand, O. (2018). The beat of the mountain: A transdisciplinary rhythmanalysis of temporal landscapes. *Landscape Research*, *44*(8), 937–951.

Gaidhani, S., Arora, L., & Sharma, B. K. (2019). Understanding the attitude of Generation Z towards workplace. *International Journal of Management, Technology and Engineering*, *9*(1), 2804–2812.

Goh, E., & Okumus, F. (2020). Avoiding the hospitality workforce bubble: Strategies to attract and retain Generation Z talent in the hospitality workforce. *Tourism Management Perspectives*, *33*, 100603.

Haddouche, H., & Salomone, C. (2018). Generation Z and the tourist experience: Tourist stories and use of social networks. *Journal of Tourism Futures*, *4*(1), 69–79.

Hall, C. M. (2019). Constructing sustainable tourism development: The 2030 agenda and the managerial ecology of sustainable tourism. *Journal of Sustainable Tourism*, *27*(7), 1044–1060.

Heape, C., & Liburd, J. (2018). Collaborative learning for sustainable tourism development. In J. Liburd & D. Edwards (Eds.), *Collaboration for sustainable tourism development* (pp. 226–243). Goodfellow Publishers.

Hovik, S., & Hongslo, E. (2017). Balancing local interests and national conservation obligations in nature protection: The case of local management boards in Norway. *Journal of Environmental Planning and Management*, *60*(4), 708–724.

Howe, N., & Strauss, W. (Eds.). (2000). *Millennials rising: The next great generation*. Random House.

Jamal, T. B., Kircher, J., & Donaldson, J. P. (2021). Re-visiting design thinking for learning and practice: Critical pedagogy, conative empathy. *Sustainability*, *13*(2), 964.

Jamal, T. B., & Menzel, C. (2009). Good actions in tourism. In J. Tribe (Ed.), *Philosophical issues in tourism* (pp. 277–243). Channel View Publications.

Liburd, J., Duedahl, E., & Heape, C. (2020). Co-designing tourism for sustainable development. *Journal of Sustainable Tourism*, *30*(10), 2298–2317.

Liburd, J., Nielsen, T. K., & Heape, C. (2017). Co-designing smart tourism. *European Journal of Tourism Research*, *17*, 28–42.

Øian, H. (2013). Wilderness tourism and the moralities of commitment: Hunting and angling as modes of engaging with the natures and animals of rural landscapes in Norway. *Journal of Rural Studies*, *32*, 177–185.

Peters, D., Loke, L., & Ahmadpour, N. (2020). Toolkits, cards and games: A review of analogue tools for collaborative ideation. *CoDesign*, *17*(4), 410–434.

Robinson, V. M., & Schänzel, H. A. (2019). A tourism inflex: Generation Z travel experiences. *Journal of Tourism Futures*, *5*(2), 127–141.

Rogal, M., & Sànchez, R. (2018). Codesigning for development. In R. B. Egenhoefer (Ed.), *Routledge handbook of sustainable design* (pp. 250–262). Routledge.

Rudiløkken. (2019). Skal presentere Dovre på reiselivsseminar. *Vigga Pluss*. www.vigga. no/nyheter/2019/11/13/Skal-presentere-Dovre-på-reiselivsseminar-20385441.ece?fbclid= IwAR27eIyuTNB9UzG03S22kN_sLl9Udgyl0-rzzbrf9lnYQV2l1h-2374obzo

Sakdiyakorn, M., Golubovskaya, M., & Solnet, D. (2021). Understanding Generation Z through collective consciousness: Impacts for hospitality work and employment. *International Journal of Hospitality Management, 94*, 102822.

Scholz, C. (2019). The Generations Z in Europe: An introduction. In C. Scholz & A. Rennig (Eds.), *Generations Z in Europe: Inputs, insights and implications* (pp. 3–31). Emerald Publishing.

Seemiller, C., & Grace, M. (Eds.). (2019). *Generation Z: A century in the making.* Routledge.

Seyfi, S., & Hall, C. M. (2021). COVID-19 pandemic, tourism and degrowth. In C. M. Hall, L. Lundmark, & J. Zhang (Eds.), *Degrowth and tourism: New perspectives on tourism entrepreneurship, destinations and policy* (pp. 287–313). Routledge.

Sharpley, R. (2020). Tourism, sustainable development and the theoretical divide: 20 years on. *Journal of Sustainable Tourism, 28*(11), 1932–1946.

Skinner, H., Sarpong, D., & White, G. R. T. (2018). Meeting the needs of the millennials and Generation Z: Gamification in tourism through geocaching. *Journal of Tourism Futures, 4*(1), 93–104.

Skogen, K., Krange, O., & Figari, H. (2017). *Wolf conflicts: A sociological study* (Vol. 1). Berghahn Books.

Su, C. H., Tsai, C. H., Chen, M. H., & Lv, W. Q. (2019). US sustainable food market Generation Z consumer segments. *Sustainability, 11*(13), 3607.

Timmermans, S., & Tavory, I. (2012). Theory construction in qualitative research: From grounded theory to abductive analysis. *Sociological Theory, 30*(3), 167–186.

Törőcsik, M., Szűcs, K., & Kehl, D. (2014). How generations think: Research on Generation Z. *Acta Universitatis Sapientiae, Communicatio, 1*(1), 23–45.

United Nations (UN). (2016). *Transforming our world: The 2030 agenda for sustainable development.* https://sdgs.un.org/2030agenda

Wee, D. (2019). Generation Z talking: Transformative experience in educational travel. *Journal of Tourism Futures, 5*(2), 157–167.

Yamane, T., & Kaneko, S. (2021). Is the younger generation a driving force toward achieving the sustainable development goals? Survey experiments. *Journal of Cleaner Production, 292*, 125932.

4 Are Generation Zers *Homo vians*, *Phono sapiens*, and *Homo ecologicus*? Intergenerational comparison with a reference to the Republic of Korea

Jungho Suh

Introduction

MacCannell (2013) put forward the notion that a tourist is an actual person and any actual person is a tourist. This remark illuminates that human beings are inherently 'travelling beings' (*Homo vians*). No wonder, people in modern society often travel to a place away from their usual environment in search of the meaning of their life (Pernecky & Poulston, 2015; Sirirat, 2019). Generation Z (Gen Z) is not exceptional. They are eager to see how other people live in other locations and learn about themselves from sightseeing.

Gen Z has grown up in the era of the digital revolution, being known as digital nomads or digital natives. They are connected to an ocean of information at any time through the Internet and social media. Most information they need to find is just a few clicks away. Gen Z is familiar with sharing travel information via various web-based platforms and application programmes in real time (Haddouche & Salomone, 2018; Styvén & Foster, 2018). For example, they can deftly find someone at short notice who is willing to share a jeep to travel from Leh Town to Nubra Valley in Ladakh, far northern India, and travel together as per the same itinerary.

On one hand, Gen Z is often understood as addicted to digital devices such as smartphones (Sutton, 2020). Addiction to Internet games and social media may result in limiting the time spent in growing their thinking, reading, and writing skills, as does easy access to the Internet and heavy reliance on Internet materials as the source of information. No wonder, the media often reports that a growing number of students are suffering from mental health problems. For this reason, Gen Z is in much need of spiritual tourism, yoga tourism, healing tourism, leisure tourism, and recreational tourism.

On the other hand, many tourism scholars (e.g., Bulut et al., 2017; Seemiller & Grace, 2019; Dabija et al., 2020) came to the view that Gen Z tends to avoid unnecessary consumption and supports social enterprises and eco-friendly retail shops. However, it is questionable whether digital natives retain truly ecocentric environmental attitudes and behaviour. As a matter of fact, travelling by fossil-fuel-burning transport modes entails considerable emission of greenhouse gases and contributes to global warming. Strong ecocentric environmentalism without synthetically analysing the causes of, and policy responses to, contemporary

DOI: 10.4324/9781003289586-6

environmental pollution is nothing short of blind environmentalism, which is not conducive to saving the troubled world.

The current tourism literature lacks in-depth discussion as to how the rather conflicting attributes – namely, *Homo vians*, *Phono sapiens*, and *Homo ecologicus* – are combined to characterise Gen Z. In a way to fill in the research gap, this chapter investigates the characteristics of those in their 20s in the Republic of Korea (hereafter Korea) with a focus on their tourism experience and environmental behaviour. It can be hypothesised that Korean young adults are not genuinely interested in environmental issues whereas they are addicted to online games played on smartphones. If this null hypothesis is rejected, one can investigate whether there is no difference between generations in terms of their environmental attitudes and behaviour. The chapter first looks into the trend of Korean outbound tourism demands from 1995 to 2020. The chapter next outlines which source they use to collect the information on their tourism destinations. The chapter then tracks down the environmental attitudes and behaviour of Korean young adults.

Research methods

This study makes use of secondary data sources including the Korea Tourism Statistical Yearbook 2020 (Korea Tourism Organization, 2020) and Korean Tourism Data Lab (Korea Tourism Organization, 2022). In these statistical data, the ages of travellers are classified into less than 20s, 20s, 30s, and so on rather than Gen Z, Y, and X. Those who are in their 20s as of 2020 include the younger members of Gen Y who were born in the years 1990 to 1994 as well as the older members of Gen Z who were born in the years 1995 to 2000.

Korea has been selected for multiple reasons. Korea has experienced rapid economic growth since the 1960s. The Korean gross domestic product (GDP) per capita was US$1,027 in 1960 (in constant 2015 US dollars) and increased to US$31,265 in 2020 (World Bank, 2022a). The United Nations Conference on Trade and Development (UNCTAD; 2022) recognised that Korea had become an economically affluent society by global standards and reclassified the country into Group B (developed economies) in 2021. It can be predicted that Korean outbound tourism has flourished, owing to fast economic growth in such a short time span.

Second, the Korean economy also underwent the Asian economic crisis in 1998 and the world economic crisis in 2008. This means that Korean Gen Zers were born in an era of economic downturns accompanied by jobless growth, high unemployment rates, and fierce competition in job markets. Consequently, they grew up in the aftermath of economic crises, where many of their parents' generation might have suffered from job losses. Thus, it can be expected that Korean Gen Zers travel overseas less often than older generations in the face of economic crises.

Third, Korean Baby Boomers born between 1946 and 1964 were in their late 20s to early 50s in the 1990s when Korea was at the peak of economic growth rates. In the 1990s, on the other hand, environmental pollution and degradation caused by economic growth began to come to public attention. The Korean public perception of environmental quality started changing in the 1990s when Gen Z came into the world (Handy et al., 2021). This means that Korean Gen Zers grew up in the face of critical environmental pollution and signs of global warming and consequent severe weather events.

Fourth, Korea is known for the high uptake of smartphones with 89% of the Korean population having a smartphone and with 137.5 registered mobile phone numbers per 100 persons in 2020 (Korean Statistical Information Service, 2022). About 92% of Korean adults use the Internet to have access to news media or social networks in everyday life (National Information Society Agency, 2021).

In order to detect whether Korean Gen Zers are interested in environmental issues and climate actions, this study has searched naver.com, the largest portal website specialising in Korean. Within the Naver domain, there are numerous conversation clubs and blogs related to travel or tourism. This study typed 'Auroville', 'Findhorn', 'Crystal Waters', and 'Svanholm' in the search field of the Naver blogs and conversation clubs. The rationale for this is multiple-fold. First, visiting an ecovillage as a tourist destination would be a suitable indicator that the visitors are interested in ecological human settlements and lifestyles (van Schyndel Kasper, 2008; Prince, 2017; Brombin, 2019; Doğan, 2019). Second, travelling to these ecovillages requires outbound tourism because there is no intentional community of this kind in Korea. Third, these ecovillages are located in countries which are popular outbound tourism destinations for young Korean backpackers (Korea Tourism Organization, 2020). This study manually canvasses and analyses the contents, themes, and hashtags of the blogs and travel writings posted by Korean young adults. The study could not use any software for document analysis purposes because no software is available for sources written in Korean.

Demand for outbound tourism in Korea

Veblen (1899/1994) characterised the upper class in the nineteenth century as having leisure time because many were occupied with non-industrial activities. MacCannell (2013) postulated that leisure was no longer a privilege of the upper class and saw tourism as a pivot of modern civilisation, and the spread of international tourism as a symbol of post-industrial society. MacCannell (2013) understood tourists as sightseers, mostly from the middle class which emerged in the 1980s. In this context, it was timely that the United Nations World Tourism Organization (2001) declared tourism, domestic or international, as a human right and one of the best possible expressions of the sustained growth of leisure time.

Outbound travel by Koreans has increased in proportion to GDP per capita as presented in Table 4.1. The number of outbound tourists increased 6.8 times from 4.2 million in 1995 to 28.7 million in 2019 while the total population of Korea increased by about 14.6% from 45.1 million in 1995 to 51.7 million in 2019 (World Bank, 2022b). The outbound tourist numbers in 2020 sharply dropped due to the outbreak of COVID-19.

There is no statistical evidence that those in their 20s as of 2020 have less overseas travel experience than other age groups. Generally speaking, the opportunities for outbound travel for this cohort are constrained by a lack of financial affluence

Table 4.1 Outbound tourists by age group in Korea from 1995 to 2020 (thousand persons)

Year	GDP per capita[a]	Total outbound travel[b]	Outbound travel by age group[b]						
			<20s	20s	30s	40s	50s	>50s	Crew
1995	13,409	4,156	298	892	1,142	826	620	377	–
			(7.2)	(21.5)	(27.5)	(19.9)	(14.9)	(9.1)	–
2000	16,992	5,795	475	1,040	1,328	1,103	699	470	680
			(8.2)	(18.0)	(22.9)	(19.0)	(12.1)	(8.1)	(11.7)
2005	21,193	10,372	1,065	1,698	2,218	2,252	1,387	881	871
			(10.3)	(16.4)	(21.4)	(21.7)	(13.4)	(8.5)	(8.4)
2010	25,451	12,488	1,318	1,985	2,588	2,445	1,923	1,161	1,069
			(10.6)	(15.9)	(20.7)	(19.6)	(15.4)	(9.3)	(8.6)
2015	28,732	19,310	2,197	3,149	3,910	3,639	3,188	1,803	1,425
			(11.4)	(16.3)	(20.3)	(18.9)	(16.5)	(9.3)	(7.4)
2019	31,611	28,714	3,545	4,842	5,415	5,214	4,734	3,171	1,794
			(12.3)	(16.9)	(18.9)	(18.2)	(16.5)	(11.0)	(6.3)
2020	31,265	4,276	558	701	647	743	625	414	587
			(13.1)	(16.4)	(15.1)	(17.4)	(14.6)	(9.7)	(13.7)

Source: [a] World Bank (2022a); [b] Korea Tourism Organization (2020).

Note:
[a] In constant 2015 US dollars.
[b] Figures in brackets are percentages. The total percentage does not add up to 100% due to rounding.

because many are likely to be students, job seekers, or early-career workers. However, the life cycle effect is not evident as far as Korean outbound travelling is concerned. The distribution of tourist numbers across age groups has not dramatically changed from 1995 through 2020. Moreover, this study has not found evidence of the period effect that macroeconomic circumstances might have influenced the trend of outbound tourism demands in Korea. For example, those who were in their 20s in 2000 accounted for 18.0% of the total outbound tourist numbers. The proportion slightly decreased to 17.4% when the same-age cohort reached their 40s in 2020.

Korean outbound tourists, regardless of age group, travel for multiple purposes. Indeed, it is hard to distinguish one type of tourist from another because a tourist can be engaged in various tourism activities and experiences (Berger, 2013; Wee, 2019). According to Korea Tourism Organization (2022), Korean outbound tourists are engaged in diverse types of tourism in one trip including nature tourism (81.2%), gastronomic tourism (56.5%), shopping tourism (33.6%), and heritage tourism (31.5%). They select tourism destinations because of sightseeing attractions (51.8%), popularity (55.9%), and budget constraints (23.8%).

Collection, analysis, and communication of tourism information

Berkeley (2015) coined the term '*Phono sapiens*' in an article entitled 'Planet of the phones', which was published in the 28 February 2015 edition of the weekly magazine *The Economist*. *Phono sapiens* are in a habit of collecting information and communicating through the Internet or social network services. Using smartphones, they take notes by typing. Older generations are accustomed to reading hard copies of published materials to collect the information they need. Older generations are more familiar with writing notes on a script of paper with a pen than typing into a smartphone. Therefore, they can be called *Peno sapiens* as opposed to *Phono sapiens*.

In Korea, those aged in their 20s and 30s (Gen Z and Millennials) are more likely to rely on the Internet than those in their 40s and 50s (Gen X and Baby Boomers). Younger generations do not need to carry a hard copy of a travel guidebook or talk to a local travel agent to get informed about their travel destinations. By contrast, 40- and 50-year-olds tend to rely on travel agents, compared to their following generations. Interestingly, those who were born in 2000 or after, tend to rely on travel agents as much as the Internet, compared to those in their 20s and 30s as indicated in Table 4.2.

Korean Gen Zers and Millennials seek information on transport, accommodation, and places to visit via online platforms such as tripadvisor.com or booking. com. They opt to analyse anonymous online reviews carefully, checking cleanliness as well as proximity to tourist hot spots. This is the case for the same age cohorts in other countries, for example, Taiwan (Chang & Wang, 2018). These age cohorts are also highly active in sharing information through various online communication platforms (Bernardi, 2018).

According to Naver (2022) in the Korean context, the total number of blog postings has amounted to 2.1 billion since 2003 when the portal site was opened. There are also a number of online travel conversation clubs established within naver.com. They are neither travel agents nor commercial platforms where the visitors come and book flight tickets. The club visitors just upload post-travel writings and like to share their travel experiences and reflections. The largest age cohorts using the Naver domain for this purpose in early 2022 were the 20- and 30-year-olds who accounted for 34.6% and 28.5% of the online travel writings, respectively.

Table 4.2 Sources of tourism information by age group in Korea in 2019

Source	Age group						
	<20s	*20s*	*30s*	*40s*	*50s*	*60s*	*>60s*
Previous experience	3.4	13.1	7.9	9.0	7.0	7.2	5.1
Guidebook	1.9	7.6	6.5	6.8	6.0	1.7	0.8
Advertisement	4.4	10.0	13.2	10.7	12.2	11.1	10.7
Media report	12.9	15.0	13.2	15.4	13.5	10.2	14.7
Travel agent	43.7	16.1	26.1	44.7	48.1	42.4	42.8
Internet/mobile app	49.5	67.4	64.6	44.9	33.3	17.0	8.2
None	0.0	1.8	1.6	4.2	3.7	6.6	4.2
Family/friends	55.7	63.4	58.1	60.3	57.8	63.2	73.7
Other	1.9	0.1	0.0	0.0	0.1	0.2	0.0

Source: Korea Tourism Organization (2022).

Tourism, ecovillages, and travelogues

Yes24 (2022), one of the largest online bookshops in Korea, reported that books on ecology and the environment have increasingly been sold in 2018, 2019, and 2020 with growth rates of 10.5%, 24.3%, and 217.5%, respectively. The sharp increase in 2020 was due to the outbreak of the COVID-19 pandemic when people were unable to travel or shop face to face. According to Yes24 (2022), 20- and 30-year-olds have driven the increase in the growth of book sales. The environment-related bestselling books in Korea include *Silent spring* (Carson, 1962/2011), *Small is beautiful* (Schumacher, 1973/2002), *Ancient futures* (Norberg-Hodge, 1991/2015), and *Corona sapiens* (Choi et al., 2020).

Listed among other bestselling books on environmental subjects are those books that introduce ecovillages around the world, for example, Auroville in India, Findhorn in Scotland, Crystal Waters in Australia, and Svanholm in Demark. These are intentional communities that were formed under the banner of taking action on sustainable human settlements (Litfin, 2014). They have been marching towards making a place with social cohesion, low-impact food production, and energy-saving housing, moving away from the anthropocentric and individualistic capitalist mindset. The books on ecovillages published in Korean have induced outbound travelling to ecovillages amongst Korean young adults. Many of those who had already visited the ecovillages have published travel blogs, which have in turn led to a phenomenal increment in tourism to ecovillages amongst other Korean young adults.

Tying in each of 'Auroville', 'Findhorn', 'Crystal Waters' and 'Svanholm' in Korean, this study hit a total of about 5,140 travel blogs and online discussion threads posted between 2010 and 2021 to naver.com, as presented in Table 4.3. The contributors wrote about their experiences and impressions of the ecovillages they visited. This study analyses the contents and themes of these entries posted by those in their 20s and 30s as of 2021, who had ever visited the communities as their primary or one of the main legs of their tourism destinations. An overwhelming majority of the postings related to Auroville and Findhorn, which are known as spiritual intentional communities.

Auroville was more popular than Findhorn thanks to a diverse range of voluntourism and experience programmes. International tourists to Auroville are generally short-term overnight visitors or volunteer workers. Sadhana Forest, an Auroville working group that focuses on greening projects, is well known to Korean volunteer tourists. It was founded in 2003 by Aviram and Yorit Rozin. They have since reforested 28 hectares (ha) of severely eroded and infertile land with

Table 4.3 Number of postings uploaded to Naver (2010–2021)

Naver	Auroville	Findhorn	Crystal Waters	Svanholm
Conversation club	713	345	142	57
Travel blog	2,188	832	508	355

indigenous tropical evergreen tree species (Prisma, 2011). Sadhana Forest produces electricity with solar panels and sticks to veganism. According to the Naver blogs, the Korean volunteer workers were astonished that there was no flush toilet in Sadhana Forest and many compost toilets were put in place to save water and to recycle into natural fertilisers. The Korean volunteer workers were proud of making their contribution to the Sadhana reforestation and forest management projects.

Many Korean travel bloggers who had been to Auroville commonly expressed they were impressed with the institutional arrangement of the communitarian sharing economy being implemented in Auroville. Under this scheme, adult Aurovillians are expected to contribute to the collective welfare according to their capacities in the form of labour or money (Thomas & Thomas, 2013). Aurovillians who contribute their labour through a working group, a service unit, or a business unit may receive a monthly maintenance allowance, which is meant to meet their basic living costs.

The hashtags of the online travel writings indicate that Korean youths are interested in environmental issues in the broader and down-to-earth context of human ecology, including rural return migration, organic farming, resource recycling, social enterprises, and transition towns. Korean young adults are familiar with the rural return migration movement as Korean local governments have competitively and aggressively promoted rural-urban youth migration to rejuvenate declining rural communities. The concept of transition towns is also well known to Korean young adults, thanks to the Korean version of urban retrofitting programmes initiated by the former Seoul Mayor.

The Korean travel bloggers pointed out that the ecovillages they visited were not a 'utopia' or a 'paradise' and not without constraints and limitations. For instance, only a small proportion of staple food was produced within the communities. The steady development of Auroville has increased the price of land in the vicinity and triggered real estate speculation. This situation has made it difficult for Auroville to acquire the land needed to turn the outer circle of the township into a complete greenbelt as planned. In addition, traffic guides at major intersections in Auroville have to wear a mask due to massive dust from unpaved roads and emissions from automobiles. Although disappointed with all the shortcomings, the Korean young travellers to the aforementioned ecovillages became determined to continue to look for a viable alternative to growth-driven capitalist economic and social systems.

Discussion

This study rejects the hypothesis that Korean young adults are not genuinely concerned about environmental pollution and degradation while they are addicted to Internet surfing and online games. Korean Gen Z has witnessed and experienced unprecedentedly high youth unemployment rates due to economic downturns worldwide, which have been exacerbated by the COVID-19 pandemic and consequent travel restrictions. This age cohort is conscious of local and global environmental pollution, which is attributed to a consumerism lifestyle, and competition-based profit-maximising business behaviour and practices. By nature, they advocate grassroots environmental stewardship actions towards a localised, circular, and sharing economy.

The Korean cohort aged in their 20s is interested in the mechanism of a sharing economy to the point where they are keen to travel overseas to visit ecovillages. Many ecovillages (e.g., Auroville, Findhorn, Crystal Waters, Svanholm) work with the World Wide Opportunities on Organic Farms (WWOOF) network and offer agricultural experience programmes, which attract a wide range of mindful and responsible Gen Z tourists. Travelling to ecovillages is a form of new-age tourism, as opposed to conventional mass tourism, in the context that the tourists identify the destination attractions on their own, value ecological sustainability, and explore an alternative lifestyle (Pernecky & Poulston, 2015).

Auroville has been one of the most popular tourist destinations of its kind and is on the bucket lists for many Korean youths who were inspired by travel blogs posted to Naver.com. Analysing the online travelogues, this study has found that the Korean youth visitors to Auroville interacted with other visitors and took part in a wide range of Auroville activities. Whatever their initial and primary motivations were, the Korean youths ended up expanding the scope of their interests and experiences. They were impressed that a sharing economy was not just an idea but was taking place in the Auroville community. They were learning that a sharing economy did not necessarily jeopardise personal freedoms and that individual self-interest can be balanced with the common interest of the community, as postulated in Botsman and Rogers (2010).

From analysing the travel writings, this study finds that experiences and learnings by the visitors to the intentional communities are moving towards the realisation that spiritual life, ecological resilience, human unity, and collaborative production activities are permeating one another. The interrelatedness of apparently separate elements is illustrated in Figure 4.1. It is of great importance that the concept of 'interdependence' is different from that of 'balancing' when it comes to environmentalism or sustainability views. An analogy is that all the factors

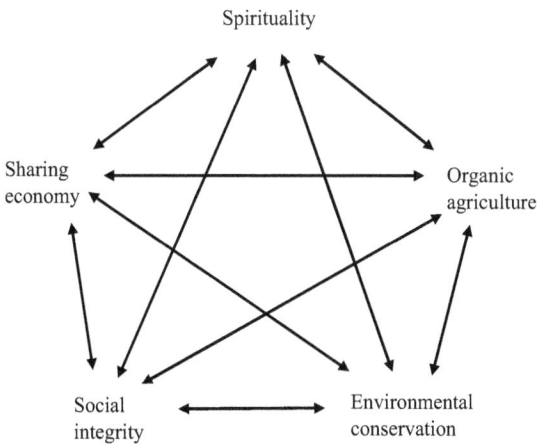

Figure 4.1 Interdependence between the spiritual, agri-environmental, social, and economic aspects of sustainability

in the IPAT equation (Impact on the environment = Population × Affluence × Technology) are interrelated (Chertow, 2001). Litfin (2014) pointed out that the ecological, economic, social, and spiritual domains of sustainability are not for balancing out but for reflecting and synergising one another. At odds, the current sustainability literature is still dominated by dualistic thinking, which assumes that sustainability elements are disparate independent variables. For example, environmental impact analysis takes it for granted that the environmental, economic, and social impacts of a development project can be separately measured for decision-support purposes. A matrix of sustainability indicators is often employed to measure the magnitude of community sustainability. For instance, Liu (2018) attempted to develop a new sustainability index that embraces economic, environmental, and social aspects of sustainability, to assess the overall sustainability of eco-towns. However, the matrix approach to measuring overall sustainability undermines the very spirit of eco-towns – the interdependence of sustainability pillars.

Despite the ubiquitous environmental concerns and actions amongst the cohort aged in their 20s, this study does not support the presumption that Korean young adults are more environmentally concerned than older generations. First, those in their 30s and 40s also upload their travel writings to social network services including Naver. The popular tourism information platforms in Korea were created and managed by Millennials. Second, the role of education played by older generations for young generations should not be neglected. Baby Boomers in Korea, who were born before 1965 in one of the poorest countries in the world, were motivated to save and recycle resources out of economic necessity and frugality (Handy et al., 2021).

It is interesting to observe an environmental Kuznets curve relationship in which Korean public awareness has been growing as GDP per capita income expands and time has passed. In the early years of economic development, the public regarded environmental pollution as an unavoidable by-product of economic growth or as the price that Korea had to pay to join the world of developed economies. A survey conducted in 1984 by the Institute of Social Sciences at the Seoul National University (Shin, 1993) revealed that 75% of the interviewees regarded economic growth as 'somewhat' or 'very' important while 25% considered environmental issues of urban air, water, and noise as being 'somewhat' or 'very' important. The public perception of environmental pollution in Korea had significantly changed in the 1990s: About 87% of the respondents in a survey affirmed they were ready to endure hardships to mitigate environmental pollution and about 93% of the survey respondents viewed polluting the environment as a crime (Yoo, 1997). Media coverage frequency may reflect environmental consciousness among the public. Only a limited number of weekly 'environment' sections were allotted in several daily newspapers back in the 1980s. In 2021, countless environment-related incidents and issues are reported every day by a much higher number of news media, ranging from fine particulate matter and solid municipal waste to water pollution and carbon emissions.

Concluding comments

Korea's Gen Z is active in producing and sharing value-added information through Internet blogs, online communities, and personal homepages. They directly communicate with the hosts of accommodation. With the development and wide spread of information and communication technology, they can post-travel writings to Facebook, blogs, and other platforms. Not only are Gen Zers *Phono sapiens*, but they are also avid readers of books on ecological and environmental issues. They are strongly interested in practising what they learn from reading to combat environmental pollution rather than just talking about what can be done. For example, they are active in participating in local sharing-economy programmes. Beautiful Store (2019) reported that young people under 30 accounted for 39% of the participants in Beautiful Store, a Korean non-governmental organisation, which runs a second-hand-goods sharing programme nationwide. They donate used goods such as books and clothes to be sold at Beautiful Store or purchase second-hand goods from the store.

Unsurprisingly, Korean Gen Zers are found to be active in carrying out volunteer tourism to intentional communities to experience a sharing economy. During their stay in ecovillages, for example, Auroville and Findhorn, the young tourists are keen to learn what it is to live in a community where the spiritual, ecological, social, and economic dimensions of sustainability are simultaneously and collectively pursued and practised. Tourism to these spiritual ecovillages can be characterised by a taxonomy of tourism including spiritual tourism, yoga tourism, leisure tourism, cultural tourism, ecotourism, volunteer tourism, and alternative tourism, which all overlap one way or another. This premise echoes the point made by Wee (2019) and Su et al. (2019) that the dichotomies of tourism and tourist typologies need to be revisited because various types of tourism can occur as a part of the same trip.

Gen Z is no less interested in practising a sustainable lifestyle and travelling to sustainable communities than older generations. Nevertheless, that does not mean Gen Z is more environmentally aware than older generations. The generations preceding Gen Z also read books on environmental subjects and are concerned about the environmental consequences of the growth-oriented economic system. Some of them also visit ecovillages with their children for educational purposes. The characteristics of young Koreans identified in this study may not be representative of the entire population of Korean Gen Z and cannot be generalised across other countries. Thus, the prepositions in Seemiller and Grace (2019, p. xix) remain unchallenged: "Along with elasticity between generational cohorts, there is also some variance within generations".

References

Beautiful Store. (2019). *Statistical yearbook of beautiful store 2018* (in Korean). Beautiful Store.

Berger, A. A. (2013). *Theorizing tourism: Analysing iconic destinations*. Routledge.

Berkeley, J. (2015). Planet of the phones. *The Economist*. www.economist.com/leaders/2015/02/26/planet-of-the-phones

Bernardi, M. (2018). Millennials, sharing economy and tourism: The case of Seoul. *Journal of Tourism Futures*, *4*(1), 43–56.

Botsman, R., & Rogers, R. (Eds.). (2010). *What's mine is yours: The rise of collaborative consumption*. Collins.

Brombin, A. (2019). The ecovillage movement: New ways to experience nature. *Environmental Values*, *28*(2), 191–210.

Bulut, Z. A., Çımrin, F. K., & Doğan, O. (2017). Gender, generation and sustainable consumption: Exploring the behaviour of consumers from Izmir, Turkey. *International Journal of Consumer Studie*s, *41*(6), 597–604.

Carson, R. (1962/2011). *Silent Spring* (E. Kim, Trans., rev. ed.). Mariner Books.

Chang, W. L., & Wang, J. Y. (2018). Mine is yours? Using sentiment analysis to explore the degree of risk in the sharing economy. *Electronic Commerce Research and Applications*, *28*, 141–158.

Chertow, M. R. (2001). The IPAT equation and its variants: Changing views of technology and environmental impact. *Journal of Industrial Ecology*, *4*(4), 13–29.

Choi, J., Chang, H., Choi, J. B., Hong, K., & Kim, N. (2020). *Corona sapiens* (in Korean). Influential.

Dabija, D. C., Bejan, B. M., & Puşcaş, C. (2020). A qualitative approach to the sustainable orientation of Generation Z in retail: The case of Romania. *Journal of Risk and Financial Management*, *13*(7), 152.

Doğan, M. (2019). Ecological ideals, sustainable tourism and the heritage concept of an eco-village: The case of Arcosanti, USA. *Journal of Heritage Tourism*, *14*(4), 371–381.

Haddouche, H., & Salomone, C. (2018). Generation Z and the tourist experience: Tourist stories and use of social networks. *Journal of Tourism Futures*, *4*(1), 69–79.

Handy, F., Katz-Gerro, T., Greenspan, I., & Vered, Y. (2021). Intergenerational disenchantment? Environmental behaviors and motivations across generations in South Korea. *Geoforum*, *121*, 53–64.

Korean Statistical Information Service. (2022). *The number of mobile phone registrations*. https://kosis.kr/statHtml/statHtml.do?orgId=101&tblId=DT_2KAAA10

Korea Tourism Organization. (2020). *Korean tourism statistical yearbook 2019*. https://data.worldbank.org/indicator/ST.INT.ARVL?locations=KR

Korea Tourism Organization. (2022). *Korean tourism data lab*. https://datalab.visitkorea.or.kr/datalab/portal/main/getMainForm.do

Litfin, K. T. (2014). *Ecovillages: Lessons for sustainable community*. Polity Press.

Liu, L. (2018). A sustainability index with attention to environmental justice for eco-city classification and assessment. *Ecological Indicators*, *85*, 904–914.

MacCannell, D. (2013). *The tourist: A new theory of the leisure class*. University of California Press.

National Information Society Agency. (2021). *2020 Survey on the internet usage*. www.nia.or.kr/site/nia_kor/ex/bbs/View.do?cbIdx=99870&bcIdx=23213&parentSeq=23213

Naver. (2022). *Naver blog report* (in Korean). https://campaign.naver.com/2020blog/blogreport/

Norberg-Hodge, H. (1991/2015). *Ancient futures: Learning from Ladakh* (H. Yang, Trans., rev. ed.). Sierra Club Books.

Pernecky, T., & Poulston, J. (2015). Prospects and challenges in the study of new age tourism: A critical commentary. *Tourism Analysis*, *20*(6), 705–717.

Prince, S. (2017). Working towards sincere encounters in volunteer tourism: An ethnographic examination of key management issues at a Nordic eco-village. *Journal of Sustainable Tourism*, *25*(11), 1617–1632.

Prisma. (2011). *Auroville farms, forest, and botanical gardens*. Prisma.

Schumacher, E. F. (1973/2002). *Small is beautiful: Economics as if people mattered* (S. Lee, Trans., rev. ed.). Harper and Row.

Seemiller, C., & Grace, M. (Eds.). (2019). *Generation Z: A century in the making*. Routledge.

Shin, D. H. (1993). Economic growth and environmental problems in South Korea: The role of the government. In M. C. Howard (Ed.), *Asia's environmental crisis* (pp. 235–256). Westview Press.

Sirirat, P. (2019). Spiritual tourism as a tool for sustainability: A case study of Nakhon Phanom Province, Thailand. *International Journal of Religious Tourism and Pilgrimage*, *7*(3), 97–111.

Styvén, M. E., & Foster, T. (2018). Who am I if you can't see me? The "self" of young travellers as driver of eWOM in social media. *Journal of Tourism Futures*, *4*(1), 80–92.

Su, C. H. J., Tsai, C. H. K., Chen, M. H., & Lv, W. Q. (2019). US sustainable food market Generation Z consumer segments. *Sustainability*, *11*(13), 3607.

Sutton, T. (2020). Digital harm and addiction: An anthropological view. *Anthropology Today*, *36*(1), 17–22.

Thomas, H., & Thomas, M. (2013). *Economics of people and earth: The Auroville case 1968–2008*. Prisma.

United Nations Conference on Trade and Development (UNCTAD). (2022). *Country classification*. https://unctadstat.unctad.org/EN/Classifications.html

United Nations World Tourism Organization (UNWTO). (2001). *Global code of ethics for tourism*. www.unwto.org/global-code-of-ethics-for-tourism

van Schyndel Kasper, D. (2008). Redefining community in the ecovillage. *Human Ecology Review*, *15*, 12–24.

Veblen, T. (1899/1994). *The theory of the leisure class*. Dover Publications.

Wee, D. (2019). Generation Z talking: Transformative experience in educational travel. *Journal of Tourism Futures*, *5*(2), 157–167.

World Bank. (2022a). *GDP per capita (constant 2015 US$)*. https://data.worldbank.org/indicator/NY.GDP.PCAP.KD

World Bank. (2022b). *Population, total*. https://data.worldbank.org/indicator/SP.POP.TOTL

Yes24. (2022). *An increasing number of readers have purchased books on ecology and the environment* (in Korean). http://ch.yes24.com/Article/View/44922

Yoo, C. M. (1997). Koreans dissatisfied with government environmental policy: Survey. *Korea Herald*, p. 3.

5 Leading the sustainability change? Gen Z business students navigating amid global disruptions

Miia Grénman, Juulia Räikkönen,
Outi Uusitalo, and Fanny Aapio

Introduction

We live amid an ecological crisis – the combination of accelerating climate change and biodiversity loss. Environmental problems have become major topics of discussion around the globe, suggesting that human activities play a significant role in degrading nature (Díaz et al., 2019; Jayasinghe & Darner, 2020). Simultaneously, the COVID-19 pandemic is undoubtedly one of the most disruptive events faced by humankind in recent times and has critically changed our everyday lives, practices, consumption, and tourism behaviour (Gössling et al., 2020). These global crises are unprecedented in their level of disruption and call for positive change (Hall et al., 2020; Rosenbloom & Markard, 2020).

Societal transformation as a catalyst of positive change has become topical in academic discussions in recent years (O'Brien, 2018) and refers to transforming values and behaviours and shifting the prevailing sociocultural, political, and economic paradigms towards globally responsible consumer behaviour, leadership, and enhanced visions of the 'good life' (Amel et al., 2017; Dasgupta, 2021; Díaz et al., 2019). Although marketing and consumer research has frequently examined sustainable consumption and consumer behaviour (Thøgersen & Schrader, 2012), it has inadequately addressed societal transformation and visions of the good life considering global disruptions.

While the environmental limits to economic growth have been acknowledged since the 1970s, the current power-wielding generations have been slow to effectively counter the environmentally damaging trajectories (Dasgupta, 2021; Díaz et al., 2019). However, the new generation, dubbed Generation Z (Gen Z), born between the late 1990s and the late 2000s (White, 2017), seems poised to disrupt the old ways of life, leadership, and consumption. Viewed as a beacon of hope for a more sustainable future, Gen Z comes equipped with notably different values, attitudes, beliefs, expectations, and behaviours than previous generations (Corey & Grace, 2019). Having been born into the digital era and growing up with increasing environmental consciousness, Gen Z takes sustainability seriously due to having a global mindset with ethical sensitivities (Corey & Grace, 2019).

This chapter examines how the ecological crisis and COVID-19 pandemic have affected Gen Z business students' environmental world view, environmental

DOI: 10.4324/9781003289586-7

education level, and environmental behaviour concerning sustainable everyday practices and tourism behaviour. The chapter's contribution emphasises the possibility of converting global disruptions into opportunities for learning and growth, which may lead to a positive shift and enhanced visions of the good life. Considering Gen Zers' sustainability ethos, the chapter questions whether Gen Z could lead the next 'sustainability' change. The chapter concludes that catalysing positive change in societal transformation requires solidifying Gen Zers' existing values, attitudes, and behaviours into concrete actions to be undertaken by tomorrow's consumers, professionals, and leaders.

This chapter is structured as follows: First, negative disruptions and societal transformation as catalysts for positive change are presented. Second, the environmental world view, environmental education, and environmental behaviour are briefly addressed to interpret and understand societal transformation. Third, research data and methodology are presented, after which research findings are discussed. Finally, the summary and conclusions are addressed with major implications of the chapter.

Turning negative disruptions into positive change

The world is at a turning point, approaching a significant disruption – a moment when (Gilding, 2011; Muff, 2013) "both mother nature and father greed hit the wall at once" (Friedman, 2009). Fundamentally, disruption is an event that changes a system's ongoing trajectories and prevents something from continuing as usual or expected (Cambridge Dictionary Online, 2022). The ecological crisis is a significant negative disruption altering the normal planetary functions and traditional pursuit of economic wealth and prosperity. Yet, negative disruptions can also offer hope for positive change; disruption can be harnessed to accelerate transforming technologies and economies for sustainable development (Rosenbloom & Markard, 2020; Schipper et al., 2021).

The ecological crisis calls for positive change – a force that can fundamentally alter the negative path. Societal transformation as a catalyst for positive change involves transforming values, beliefs, world views, and knowledge; the systems and structures, sociocultural, political, and economic relations; and technologies, practices, and behaviours contributing to the ecological crisis (O'Brien, 2018; Schipper et al., 2021). According to O'Brien (2018), societal transformation can occur in three embedded and interacting spheres: personal (values and world views); political (systems and structures); and practical (technologies and behaviour). Individual and collective values and world views shape how the systems and structures are viewed and influence what types of technologies and behaviour are considered possible to achieve positive change. The chapter discusses societal transformation through the environmental world view, environmental education, and environmental behaviour.

Environmental world view, education, and behaviour

The concept of environmental sensitivity arose in industrialised countries over 30 years ago, and its importance has further increased due to the accelerated ecological crisis. Environmental sensitivity refers to an individual's concern, respect, and empathy for the environment (Cheng & Wu, 2015). Sensitivity towards the environment is developed from experiences with nature and time spent in natural surroundings, positively correlating with an individual's relationship with nature (Kukkonen et al., 2018; Kyriakopoulos et al., 2020).

Different scales have been developed to measure environmental sensitivity, of which one of the most extensively used is the New Ecological Paradigm (NEP) scale (Dunlap et al., 2000) measuring an individual's environmental world view. The NEP scale includes various statements related to ecological limits to growth, anti-anthropocentrism, the fragility of nature's balance, rejection of human exceptionalism, and belief in eco-crisis (Kukkonen et al., 2018).

As learning about environmental sensitivity is a lifelong and hierarchic process, environmental sensitivity should be promoted before conveying environmental knowledge (Kukkonen et al., 2018). Environmental education includes approaches, tools, and programmes developing and supporting environmentally related values, attitudes, awareness, knowledge, and skills preparing individuals to take informed action on the environment's behalf (Ardoin et al., 2020). Instead of a linear path from environmental sensitivity to knowledge to action, environmental education is understood as a dynamic and complex process influencing behaviour in multiple ways (Ardoin et al., 2020).

Universities produce science-based knowledge needed to advance environmental education. Previous research has portrayed universities as institutions preparing future professionals, leaders, decision-makers, and scholars (Ferrer-Balas et al., 2010; Kyriakopoulos et al., 2020). However, recent literature has also acknowledged the importance of business schools in shaping business ethics and corporate social responsibility (Delgado et al., 2020; Novo-Corti et al., 2018) and environmental education and management (Jabbour, 2010; Kyriakopoulos et al., 2020; Suárez-Perales et al., 2021).

Although it is acknowledged that business schools must take a leadership position in addressing the ecological crisis (cf. Jabbour, 2010), research incorporating ecological issues and environmental management into business school activities is still relatively scarce (Kyriakopoulos et al., 2020). Muff (2013) has argued that if business schools wish to positively transform business and society, they must embrace a significant transformation by refocusing education to ensure education globally responsible leaders, transforming research into an applied field, and enabling business organisations to serve the common good and engage in transforming business and the economy by actively participating in the ongoing public debate.

By developing business students' environmental knowledge, students are likely to become more concerned about the environment and motivated to engage in pro-environmental behaviour (Cheng & Wu, 2015; Kukkonen et al., 2018). As people continuously interact with their environment, nearly all human behaviour could be called environmental behaviour. Previous research has suggested that increasing individuals' environmental knowledge often results in more positive attitudes towards the environment and more responsible environmental behaviour, yet, advancing mere knowledge is insufficient for achieving change as behaviour stems from various factors, such as beliefs, attitudes, situational opportunities, and barriers (Cheng & Wu, 2015; Dunlap et al., 2000; Kukkonen et al., 2018).

Methodology

To gain an in-depth understanding of Gen Z business students' environmental world view, environmental education, and environmental behaviour, both quantitative and qualitative data were collected at one Finnish Business School in 2020 and 2021 as part of a basic-level course on business ethics. The survey included closed- and open-ended questions on the following themes: 1) concern about global disruptions; 2) environmental world view; 3) environmental knowledge; 4) integration of global disruptions in the business school curriculum; 5) sustainable everyday practices and consumption habits; and 6) sustainable tourism behaviour (Table 5.1).

The survey was administered via Qualtrics and completed by 172 business students aged 19 to 24. Most respondents were female (65%; male 35%). The closed-ended responses were analysed through descriptive methods with SPSS28 and open-ended responses with content analysis facilitated by NVivo20. Also, written narratives (n=70) were collected from the same business students. The students were advised to write short (200–400 words) descriptions of their tourism behaviour before, during, and after the pandemic, focusing on opinions about the justification of tourism, flight or travel shame, and future travel behaviour (Table 5.1). The narratives were also analysed with content analysis facilitated by NVivo20.

Notably, all participants were guaranteed anonymity and asked to provide written consent granting permission to utilise their responses for research. Research participation was voluntary yet encouraged for class credit.

Findings

Concern about global disruptions

Gen Zers' concern about global disruptions was measured with a 5-point Likert scale ranging from 1=not concerned at all to 5=extremely concerned. Although data collection occurred during the COVID-19 pandemic, only one-third of respondents were extremely or moderately worried about the current or future pandemics (M=3.05; SD=0.94). Regarding the ecological crisis, 75% of respondents were extremely or moderately concerned about climate change (M=4.02; SD=0.92), while 60% indicated extreme or moderate concern about the biodiversity crisis (M=3.70; SD=0.91).

Respondents were further asked to describe whether COVID-19 had altered their concern about the ecological crisis. Approximately one-third of respondents stated that their concern had increased, while most reported no change caused by

Table 5.1 Research design and operationalisation

Theme	Questions	Operationalisation
Global disruptions		
Concern about global disruptions	How concerned are you about climate change, the biodiversity crisis, and current or future pandemics?	3 closed-ended questions with a 5-point Likert scale: 1=not concerned at all to 5=extremely concerned
	Has the pandemic changed your concern about climate change and the biodiversity crisis? If so, how?	1 open-ended question
Environmental world view		
Environmental world view	How would you evaluate the following statements related to the state of the environment?	15 closed-ended questions with a 5-point Likert scale: 1=strongly disagree to 5=strongly agree (NEP scale adapted from Dunlap et al., 2000; Kyriakopoulos et al., 2020)
Environmental education		
Environmental knowledge	How would you describe your current knowledge of environmental issues and global ecological challenges?	1 closed-ended question with a 5-point Likert scale: 1=very poor to 5=very good (adapted from Kyriakopoulos et al., 2020)
Integration of global disruptions in the business school curriculum	How well do you believe your business school has incorporated global challenges into the curriculum: 1) climate change, 2) the biodiversity crisis, and 3) current and future pandemics?	1 closed-ended question with a 5-point Likert scale: 1=very poor to 5=very well (adapted from Kyriakopoulos et al., 2020)
Environmental behaviour		
Sustainable everyday practices and consumption habits	How would you evaluate the following statements related to sustainable everyday practices and consumption habits?	13 closed-ended questions with a 5-point Likert scale: 1=strongly disagree to 5=strongly agree (Environmental Behaviour [EB] scale adapted from Kyriakopoulos et al., 2020)
	Has the pandemic made your everyday practices and consumption habits more or less sustainable? If so, how?	1 open-ended question
Sustainable tourism behaviour	Do you consider tourism and travel a human right in normal circumstances? What about during the pandemic?	Written narratives
	Have you felt flight shame or more comprehensive travel shame in normal circumstances, during the pandemic, or both? If yes to any of the aforementioned questions, how many flight and travel shame affect your post-pandemic tourism behaviour?	

the pandemic. Those who conveyed increased concern seemed to associate their fears with an overall worry about global crises and their interconnectedness: *When the pandemic broke, I watched a TED Talk in which an expert explained that the more we destroy the environment, the more likely new pandemics are.*

Many also expressed that the increasing concern had simultaneously advanced their overall environmental awareness and knowledge of the ecological crisis: *The concern over climate change and the biodiversity crisis has also increased as my awareness has grown during the pandemic.*

Interestingly, those who reported that COVID-19 had not increased their worry about the ecological crisis identified various positive outcomes of the pandemic: It forced students to adopt more sustainable consumption, such as decreasing travelling and increasing online meetings: *The pandemic has essentially given me hope. It has changed our way of living and facilitated future changes in everyday practices.*

Many also stated that the pandemic had made them reconsider their consumption habits, which was seen as an asset in addressing the ecological crisis: *Due to the pandemic, many have thought about their own consumption, which is why emissions have dropped. So, it would be good to learn to apply these tools in normal life as well.*

Environmental world view

Gen Zers' environmental world view was measured with the NEP scale, including 15 statements related to environmental limits to growth, anti-anthropocentrism, the fragility of nature's balance, rejection of human exceptionalism, and belief in eco-crisis (Dunlap et al., 2000). In general, high scores on the NEP scale mean that the respondents have pro-environmental orientation and ecological awareness, which is expected to lead to pro-environmental beliefs and attitudes.

The mean values for the items ranged from 1.66 to 4.68 on a 5-point Likert scale from 1=strongly disagree to 5=strongly agree (Figure 5.1). Items 11–15 were reversed, suggesting a human-centred or anthropocentric view towards nature; thus, a low mean (M=1.66 to 2.79) on these items represents higher environmental sensitivity. The low values on these items and a high mean value on the item 'Humans are seriously abusing the environment' (M=4.68; SD=0.64) indicate heightened awareness of the ecological crisis and disagreement with the view that humans should dominate nature.

Most respondents seemed to acknowledge that with the current trajectories, the ecological limits are approaching fast – highlighted by high mean values on the items 'If things continue on their present course, we will soon experience a major ecological catastrophe' (M=4.36; SD=0.74) and 'We are approaching the limit of the number of people the Earth can support' (M=4.31; SD=0.80). By contrast, the high mean value on the item 'When humans interfere with nature, it often produces disastrous consequences' (M=4.31; SD=0.72) relates to a strong awareness of the balance of nature.

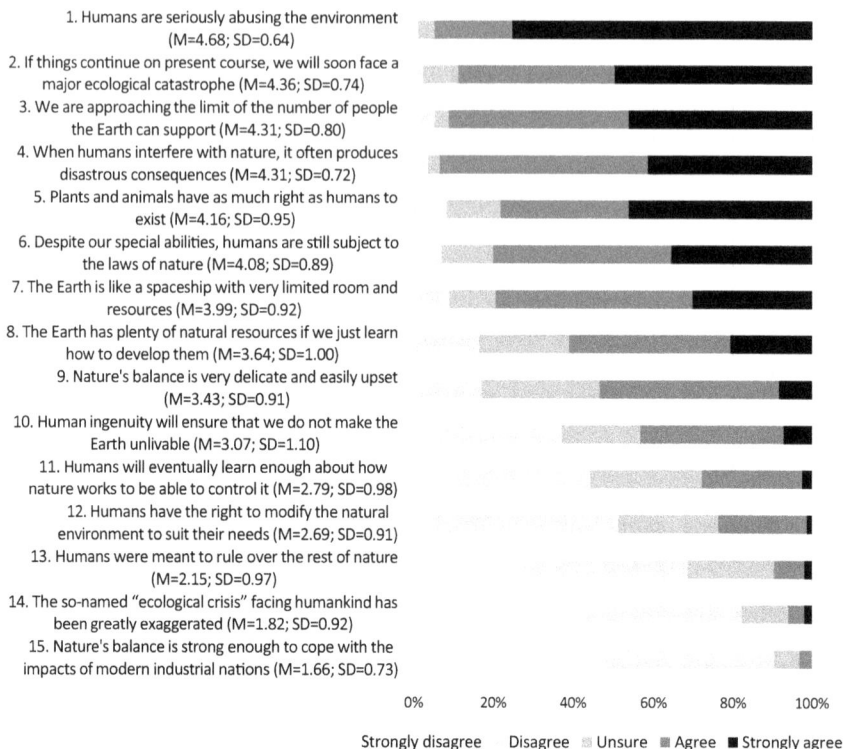

1. Humans are seriously abusing the environment (M=4.68; SD=0.64)
2. If things continue on present course, we will soon face a major ecological catastrophe (M=4.36; SD=0.74)
3. We are approaching the limit of the number of people the Earth can support (M=4.31; SD=0.80)
4. When humans interfere with nature, it often produces disastrous consequences (M=4.31; SD=0.72)
5. Plants and animals have as much right as humans to exist (M=4.16; SD=0.95)
6. Despite our special abilities, humans are still subject to the laws of nature (M=4.08; SD=0.89)
7. The Earth is like a spaceship with very limited room and resources (M=3.99; SD=0.92)
8. The Earth has plenty of natural resources if we just learn how to develop them (M=3.64; SD=1.00)
9. Nature's balance is very delicate and easily upset (M=3.43; SD=0.91)
10. Human ingenuity will ensure that we do not make the Earth unlivable (M=3.07; SD=1.10)
11. Humans will eventually learn enough about how nature works to be able to control it (M=2.79; SD=0.98)
12. Humans have the right to modify the natural environment to suit their needs (M=2.69; SD=0.91)
13. Humans were meant to rule over the rest of nature (M=2.15; SD=0.97)
14. The so-named "ecological crisis" facing humankind has been greatly exaggerated (M=1.82; SD=0.92)
15. Nature's balance is strong enough to cope with the impacts of modern industrial nations (M=1.66; SD=0.73)

0%   20%   40%   60%   80%   100%

Strongly disagree   Disagree   Unsure   Agree   Strongly agree

Figure 5.1 Gen Z business students' environmental world view according to the New Ecological Paradigm (NEP) scale

Environmental education

Concerning the level of environmental knowledge, Gen Zers described their current understanding of environmental issues and global ecological challenges to be above average (M=3.57; SD=0.64) on a 5-point Likert scale from 1=very poor to 5=very good. Over 90% of respondents reported having moderate or good environmental knowledge, while 6% stated they have very good knowledge.

Students were further asked to indicate how well the business school had incorporated the global challenges of climate change, the biodiversity crisis, and current and future pandemics in their curriculum on a 5-point Likert scale from 1=very poor to 5=very well. Respondents stated that the business school had succeeded best in integrating the current and future pandemics into the teaching (M=3.47; SD=0.89) – nearly as well as with climate-change-related issues (M=3.37; SD=0.68) – but did considerably worse with the biodiversity crisis (M=2.79; SD=0.77). More specifically, 53% of respondents felt that the business school had addressed the pandemics well or very well, while the result for climate change was 42% and only 16% for the biodiversity crisis.

Environmental behaviour

Sustainable everyday practices and consumption habits

Gen Zers' sustainable everyday practices and consumption habits were examined first through closed-ended questions via the Environmental Behaviour (EB) scale (Figure 5.2) from Kyriakopoulos et al. (2020). Second, an open-ended question was used to further clarify whether and how the COVID-19 pandemic changed their sustainable everyday practices and consumption habits. The main changes involved more attention to sustainability and increasing daily practices adhering to sustainable ideas: travelling and vehicle usage, incorporating food practices, purchasing products and services, and managing waste. Nevertheless, most respondents reported no changes due to COVID-19. These themes are discussed according to the classification by Kyriakopoulos et al. (2020): information seeking; recycling; green consumption; and active participation.

Information seeking is an essential step in behavioural change towards sustainable practices. Most respondents reported talking to others about environmental issues (item 3: M=3.57; SD=1.01) and watching TV programmes and videos about environmental problems (item 5: M=3.36; SD=1.12), implying that respondents

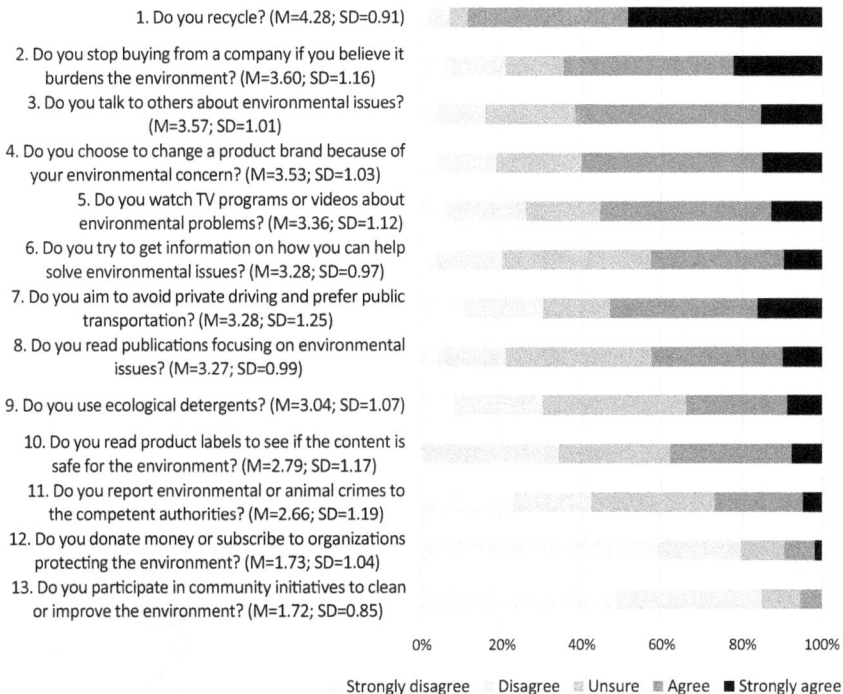

Figure 5.2 Gen Z business students' environmental behaviour according to the Environmental Behaviour (EB) scale

are keeping track of the general discussions in the media concerning the environment and society – an essential resource for sustainable transformation. Maintaining awareness and reflecting on ethical decision-making are keys to improving environmental behaviours.

Respondents were also actively seeking information about their possibilities to solve environmental problems (item 6: M=3.28; SD=0.97) or reading publications focusing on environmental issues (item 8: M=3.27; SD=0.99). The pandemic led respondents to pay more attention to products' sustainable features, thus utilising ecological information in their decision-making. Respondents described how activities during COVID-19 had brought awareness to their own practices, finding solutions to what changes are needed and how to implement them: *I have paid more attention to purchases of unnecessary products and fast fashion.*

Most respondents were also active in recycling, which was the most common sustainable practice already before the pandemic. COVID-19 had also driven some respondents to pay more attention to waste management. With more time at home, better planning, and less frequent shopping, some respondents reported more sustainable waste practices, including decreasing food waste and paying more attention to disposing of and sorting rubbish: *I have used more time for sorting waste. For example, I have donated clothes to charity and taken them to second-hand shops.*

Green consumption concerns consumption habits towards ecological products. Most respondents indicated paying attention to the consequences that products and companies have on the environment (item 2: M=3.6; SD=1.16; item 4: M=3.53; SD=1.03). However, responses to items depicting specific consumption habits, such as avoiding private driving and using ecological detergents, showed less agreement.

Regarding changing consumption practices during COVID-19, travelling and vehicle usage were the most frequently covered themes. A handful of respondents reported an increase in using private cars. While this implied that consumption became less sustainable, it was reported as a momentary response to avoid human contact in public transportation. Moreover, many described travelling less, especially on planes and reported a decrease in using private cars due to the lockdown and restrictions on activities outside the home. In many cases, cycling or walking replaced driving a car or using public transportation: *Practices are more sustainable. Travelling and consumption have decreased due to the pandemic. I have more time to follow and become conscious of what is going on in the world.*

Respondents perceived considerable barriers to shopping, preventing them from visiting stores, malls, and restaurants, leading to decreased consumption. Nonetheless, meticulousness and deliberation increased, particularly when buying new clothes. Recycling, selecting domestic brands, and disapproving of fast fashion reflected an increasing focus on sustainability. Food consumption was one theme frequently connected to sustainability changes.

As a novel daily practice, cooking at home heightened awareness and attention towards sustainable food and dining. A commonly described plot was that with home dining, vegetable consumption increased; attention was directed towards food sustainability and quality, and local or domestic food and small local restaurants were preferred: *My consumption practices have become more sustainable because time spent at home increased, which made me do things myself, such as preparing plant-based meals.*

When facing unexpected conditions due to COVID-19, some respondents took advantage of time as an essential resource enabling them to acquire knowledge and shift to green consumption practices in everyday life. Likewise, time spent at home meant that several respondents saved money, helping them pay more attention to quality, durability, and sustainability: *While staying at home, I have had time to become familiar with sustainable practices in daily life and have adopted such practices in my own life.*

The do-it-yourself capability is one resource that appeared in the data and was implemented when home-cooking allowed for more vegetable consumption, thus shifting towards more sustainable diets. More available time and changing shopping places created new opportunities: Online shopping allows browsing and searching for sustainable alternatives. 'No more hanging around in the city centre and shops' was mentioned to decrease 'unnecessary' consumption and increase sustainable practices: *As a consequence of quitting the hang-around activity in the city centre and shops, I no longer buy clothes I do not really need.*

Active participation covers commitment and participation in environmental initiatives. The responses showed a low level of involvement in this kind of environmental citizenship (item 11: M=2.66; SD=1.19; item 12: M=1.73; SD=1.04; item 13; M=1.72; SD=0.85). Gen Z business students representing well-educated consumers reported active participation in various covered attempts to make deliberate choices. One respondent expressed readiness to use education and competence to create sustainable solutions for businesses: *As a student majoring in marketing, I have attempted to come up with possibilities to create sustainable business models.*

Respondents also critically questioned whether these changes are permanent or neglected when the pandemic is over and life returns to normal. Furthermore, they speculated whether people will be determined to change their behaviour and commit to more sustainable practices in post-pandemic everyday life.

Changes in tourism attitudes and behaviour due to COVID-19

Written narratives (n=70) indicated that Gen Zers' tourism behaviour had changed significantly due to the pandemic (cf. Gössling et al., 2020; Hall et al., 2020). However, despite the main reason for restricted travel being the fear of spreading the disease, the ecological crisis was another reason for limiting international tourism.

During the pandemic, I engaged in domestic tourism only. – at first, I was afraid that I might spread the disease to small municipalities asymptomatically. – before the pandemic, I was distressed about the carbon footprint of my flights. I love to travel, and it is tough for me to prioritise the environment over my desire to explore new destinations.

Although nobody admitted travelling in secret due to the pandemic, some respondents indicated that despite travelling, they were hesitant to share pictures on social media due to peer pressure: *I'm active on social media but did not share any pictures during the first week because I worried about people's reactions.*

The narratives included value judgements (63 references) concerning tourism (Table 5.2), such as whether tourism can be considered a human right and the justification of travel restrictions during the pandemic. Some respondents stated that tourism is not a human right – neither during the pandemic nor under normal circumstances. The tourism industry was described as one of the largest polluters and not a necessity like, for example, producing food.

However, most respondents considered travelling and moving from one place to another a human right while supporting travel restrictions. Interestingly, some forms of travel and tourism were more acceptable during the pandemic than others. Business travel and visiting family abroad were considered necessary, while many saw leisure travel as unnecessary. Furthermore, some respondents judged different forms of leisure tourism: Cultural and educational tourism was considered more acceptable than beach holidays: *It is one thing to spend a week on the beach getting yourself a tan or educating yourself about places of historical and cultural significance.*

Interestingly, some respondents called tourism a privilege rather than a right and noted that only a minority of the world population could travel due to financial

Table 5.2 Gen Z business students' value judgements concerning tourism consumption

Theme	Sub-theme	Example quote
Value judgements	Tourism as an unnecessary activity	*Nobody needs tourism to live a good life. I am slightly annoyed if tourism is even mentioned with, for example, the right to education or not to be tortured.*
	Tourism as a human right	*It felt empty when I suddenly could not travel freely – like before. Travelling should be a human right, but it is unreasonable to demand access abroad during a pandemic.*
	Tourism as a privilege	*Tourism is not a human right but a privilege. Western countries are used to tourism and freedom. However, there are plenty of people who can only dream of traveling.*
Flight or travel shame	No flight or travel shame	*I have travelled a lot in my short life but have never been ashamed of it or had a bad conscience. And quite frankly, I am not ready to give it up either.*
	Felt travel or flight shame	*I have felt flight shame – as climate change and environmental issues are important to me. I was conflicted by the desire to travel and the impacts on the planet. During the pandemic, I felt relief that no one flew and there was no need to travel.*

issues or passport and visa requirements. Furthermore, tourism was described as a 'lifeline, although it is just a habit' and 'a matter of course which has become too cheap and easy'. One respondent proposed that travelling abroad has just become a way to add content to social media profiles. Another was surprised at how tourism is portrayed positively, even though many of his friends have changed other consumption habits for ethical reasons: *I hope Covid-19 has provoked people to introspect and think about the profound idea of tourism.*

Most respondents (65%) had not felt flight or travel shame (Table 5.2). The lack of guilt was explained by various reasons, such as travelling rarely or being so used to travelling. Moreover, some compared their tourism behaviour to friends or celebrities making even more unsustainable travel choices. These respondents also stated they would feel guilty if they travelled more frequently.

More than every third respondent (35%) had felt flight or travel shame due to the pandemic or tourism's environmental effects. The travel ban even seemed to relieve some of the worries over the ecological crisis. Also, some respondents described feeling shamed when travelling to their country of origin, which violated human rights, or facing human suffering in a tourist destination. In the latter, shame seemed to be combined with fear about one's safety rather than concern about local security. However, experiencing this kind of suffering was suggested to intrigue personal transformations concerning global humanity.

I flew to a big city where people are extremely rich or poor, and I was ashamed to support this kind of activity. But the experience was still valuable – it was good to see it with your own eyes and draw conclusions.

Some respondents who had felt shame justified their tourism consumption with a profound need to travel and gain experiences. In some cases, tourism destination choices and the extended length of stay were used to diminish the guilt or justify travelling despite the adverse effects: *I could not imagine going on a long-distance holiday for a week – if I fly further, I stay longer. I consider tourism choices more carefully and prefer ethical and sustainable destinations.*

Although its ecological effects were widely acknowledged, tourism was mainly seen as a positive activity. Indeed, tourism caused conflicting feelings in Gen Z: Though tourism advances climate change and is terrible for the environment, it enables them to get acquainted with the world, gain lifelong experiences, and make memories, which is hard to bargain with. According to many, no simple solutions exist: *I have begun to consider the emissions of tourism. It is a paradoxical issue; travelling has always been part of my life, but I want to learn to become a more pro-environmental consumer.*

Most respondents thought that flight or travel shame would be mostly forgotten, and tourism would eventually return to normal after the pandemic. Several respondents indicated they were unwilling to give up tourism, at least not entirely. Instead, many would rather compromise everyday consumption, such as eating vegetarian food and using public transportation. Furthermore, some were willing to compensate for their tourism and travelling by carbon offsetting their flights.

Interestingly, none questioned the effectiveness of such offsets for the environment: *I will pay more for the ticket in the future to compensate for emissions, although before the Covid-19, I have never done this.*

The minority of respondents believed that global tourism – especially international flights – would decrease in the following decades. Many were also willing to reduce tourism for the environment, while others highlighted the importance of making sustainable choices in their daily lives. Finally, some respondents indicated that the pandemic had brought about a profound transformation in the perception of tourism. While tourism was taken for granted, it has become something worth appreciating. Interestingly, some had not missed tourism as much as expected or realised travelling so much was unnecessary anyway: *The reduction in travel has been one of the positive effects of the pandemic in terms of sustainability. Travel will be thought about more carefully in the future, which is way better for the environment.*

Summary and conclusion

The ecological crisis and COVID-19 pandemic have critically changed our everyday lives, practices, consumption habits, and tourism behaviour. These global disruptions call for positive change in the direction of globally responsible consumer behaviour and leadership. In line with the widely acknowledged sustainability ethos of Gen Z, this chapter discussed Gen Z as a potential 'sustainability' change. Can the seeds of sustainable intentions embedded in Gen Z be nurtured for this generation to blossom into responsible future consumers and leaders?

The chapter addressed societal transformation, involving a fundamental change in values, world views, and behaviours, and the shift in prevailing sociocultural, political, and economic paradigms. The societal transformation was considered a catalyst for positive change and discussed through the notions of environmental world view, environmental education, and environmental behaviour. These aspects were empirically examined through quantitative and qualitative data collected from Finnish Gen Z business students.

On the question of whether Gen Z could lead the next sustainability change fuelling societal transformation, the findings are somewhat contradictory. While the study highlighted Gen Zers' strong environmental world view, knowledge, and behaviour, variation was found in the level of commitment to sustainability among Gen Zers and between everyday consumption, practices, and tourism behaviour.

The findings revealed that the students were concerned about the ecological crisis and endorsed a high level of environmental world view and knowledge. According to the previous literature, awareness, beliefs, and attitudes are antecedents to behaviour but do not necessarily indicate a strong relationship with pro-environmental behaviour due to, for example, various situational opportunities and barriers (Cheng & Wu, 2015; Kukkonen et al., 2018). Nevertheless, in the current study, the findings indicated strong pro-environmental behaviour. The students paid attention to sustainability in their daily practices, especially by seeking information, recycling, and considering their consumption choices. However, they reported a low level of commitment and participation in environmental initiatives, which is consistent with previous research (Kyriakopoulos et al., 2020).

The findings accentuated that tourism has become an essential part of Gen Zers' lives. Although the negative implications were well acknowledged, giving up tourism entirely was not seen as an option. Maintaining sustainability in everyday life seemed more reasonable than doing so in tourism, which many considered a gained advantage of breaking daily routines. However, Gen Zers also felt guilt and shame about travelling. Previous tourism literature has described youth as a crucial life stage in which travel provides transformative experiences: young travellers making memories and questioning their identities while transitioning into adulthood (Pung et al., 2020). While previous generations travelled the world with a clear conscience, that opportunity is now denied to Gen Zers navigating amid global disruptions.

The ongoing COVID-19 pandemic has changed Gen Zers' everyday consumption and tourism behaviour. Interestingly, various positive outcomes of the pandemic were identified. COVID-19 has made students reconsider their consumption habits and changed their everyday practices to be more sustainable, which was seen as an asset in addressing the ecological crisis. Moreover, the pandemic has enabled them to take advantage of time as an essential resource, enabling them to acquire knowledge and shift to even greener consumption practices. Regarding tourism behaviour, the pandemic made some consider tourism a privilege worth appreciating and others realise that frequent travel is not a necessity. Nevertheless, the students also critically questioned whether these positive changes were permanent or would be neglected when the pandemic is over and life returns to normal.

Although young generations have always been active in leading environmental movements and developments, it needs to be acknowledged that leading the change cannot be entirely Gen Zers' responsibility. This generation is facing the consequences of the global disruptions caused by previous generations and is forced to find solutions to the ecological crisis. In this task, universities and business schools have a significant role in educating Gen Zers to become a disruptor generation by catalysing the necessary change (cf. Muff, 2013). This societal transformation requires future consumers, professionals, and leaders to have moral awareness, the intention to work for sustainability, and the ability and desire to take a sustainability leadership role.

Despite the fairly small sample and narrow geographic coverage, this chapter contributes to previous consumer, tourism, and generational literature by emphasising the possibility to convert global disruptions into opportunities for learning and growth, which may lead to positive change and enhanced visions of the good life (cf. Díaz et al., 2019). Moreover, integrating positive change into discussions of the ecological crisis highlights the urgency of societal transformation. Notably, catalysing positive change requires solidifying Gen Zers' values, attitudes, and behaviours into concrete actions to be undertaken by tomorrow's consumers, professionals, and leaders, providing interesting avenues for future research.

References

Amel, E., Manning, C., Scott, B., & Koger, S. (2017). Beyond the roots of human inaction: Fostering collective effort toward ecosystem conservation. *Science*, *356*(6335), 275–279.

Ardoin, N. M., Bowers, A. W., & Gaillard, E. (2020). Environmental education outcomes for conservation: A systematic review. *Biological Conservation*, *241*, 108224.

Cambridge Dictionary Online. (2022). *Disruption*. Cambridge University Press. https://dictionary.cambridge.org/us/dictionary/english/disruption

Cheng, T. M., & Wu, H. C. (2015). How do environmental knowledge, environmental sensitivity, and place attachment affect environmentally responsible behavior? An integrated approach for sustainable island tourism. *Journal of Sustainable Tourism*, *23*(4), 557–576.

Corey, S., & Grace, M. (Eds.). (2019). *Generation Z: A century in the making*. Routledge.

Dasgupta, P. (2021). *The economics of biodiversity: The Dasgupta review*. HM Treasury.

Delgado, C., Venkatesh, M., Branco, M. C., & Silva, T. (2020). Ethics, responsibility and sustainability orientation among economics and management masters' students. *International Journal of Sustainability in Higher Education*, *21*(2), 181–199.

Díaz, S., Settele, J., Brondízio, E. S., Ngo, H. T., Agard, J., Arneth, A., Brauman, K. A., Butchart, S. H. M., Chan, K. A., Garibaldi, L. A., Ichii, K., Liu, J., Subramanian, S. M., Midgley, G. F., Miloslavich, P., Zsolt Molnár, D. O., Pfaff, A., Polasky, S., Purvis, A., . . . Zayas, C. N. (2019). Pervasive human-driven decline of life on earth points to the need for transformative change. *Science*, *366*(6471).

Dunlap, R., Liere, K. V., Mertig, A., & Jones, R. E. (2000). Measuring endorsement of the new ecological paradigm: A revised NEP scale. *Journal of Social Issues*, *56*(3), 425–442.

Ferrer-Balas, D., Lozano, R., Huisingh, D., Buckland, H., Ysern, P., & Zilahy, G. (2010). Going beyond the rhetoric: System-wide changes in universities for sustainable societies. *Journal of Cleaner Production*, *18*(7), 607–610.

Friedman, T. L. (2009, March 8). Opinion: Thomas L. Friedman: The great disruption. *The New York Times*. www.nytimes.com/2009/03/08/opinion/08iht-edfriedman.1.20672274.html

Gilding, P. (2011). *The great disruption*. Bloomsbury Publishing.

Gössling, S., Scott, D., & Hall, C. M. (2020). Pandemics, tourism and global change: A rapid assessment of COVID-19. *Journal of Sustainable Tourism*, *29*(1), 1–20.

Hall, C. M., Scott, D., & Gössling, S. (2020). Pandemics, transformations and tourism: Be careful what you wish for. *Tourism Geographies*, *22*(3), 577–598.

Jabbour, C. J. C. (2010). Greening of business schools: A systemic view. *International Journal of Sustainability in Higher Education*, *11*(1), 49–60.

Jayasinghe, I., & Darner, R. (2020). Do emotions, nature relatedness, and conservation concern influence students' evaluations of arguments about biodiversity conservation? *Interdisciplinary Journal of Environmental and Science Education*, *17*(1), e2230.

Kukkonen, J., Kärkkäinen, S., & Keinonen, T. (2018). Examining the relationships between factors influencing environmental behaviour among university students. *Sustainability*, *10*(11), 4294.

Kyriakopoulos, G., Ntanos, S., & Asonitou, S. (2020). Investigating the environmental behavior of business and accounting university students. *International Journal of Sustainability in Higher Education*, *21*(4), 819–839.

Muff, K. (2013). Developing globally responsible leaders in business schools: A vision and transformational practice for the journey ahead. *Journal of Management Development*, *32*(5), 487–507.

Novo-Corti, I., Badea, L., Tirca, D. M., & Aceleanu, M. I. (2018). A pilot study on education for sustainable development in the Romanian economic higher education. *International Journal of Sustainability in Higher Education*, *19*(4), 818–838.

O'Brien, K. (2018). Is the 1.5 C target possible? Exploring the three spheres of transformation. *Current Opinion in Environmental Sustainability*, *31*, 153–160.

Pung, J. M., Yung, R., Khoo-Lattimore, C., & Del Chiappa, G. (2020). Transformative travel experiences and gender: A double duo ethnography approach. *Current Issues in Tourism*, *23*(5), 538–558.

Rosenbloom, D., & Markard, J. (2020). A COVID-19 recovery for climate. *Science*, *368*(6490), 447.

Schipper, E. L. F., Eriksen, S. E., Fernandez Carril, L. R., Glavovic, B. C., & Shawoo, Z. (2021). Turbulent transformation: Abrupt societal disruption and climate resilient development. *Climate and Development*, *13*(6), 467–474.

Suárez-Perales, I., Valero-Gil, J., Leyva-de la Hiz, D. I., Rivera-Torres, P., & Garcés-Ayerbe, C. (2021). Educating for the future: How higher education in environmental management affects pro-environmental behaviour. *Journal of Cleaner Production*, *321*, 128972.

Thøgersen, J., & Schrader, U. (2012). From knowledge to action – new paths towards sustainable consumption. *Journal of Consumer Policy*, *35*(1), 1–5.

White, J. E. (2017). *Meet Generation Z: Understanding and reaching the new post-Christian world*. Baker Books.

6 Motivations and spatiotemporal behaviour in an urban destination

A comparative analysis between backpackers from Generations Z and Y

Márcio Ribeiro Martins and Rui Augusto da Costa

Introduction

In the last few decades, we have been witnessing fast growth in youth tourism, particularly among backpackers. This growth brings several implications for the management of tourism destinations, on how tourists consume and experience their territories, namely their space–time behaviour. The purpose of this chapter is to discuss the motivations and sustainable spatiotemporal behaviour of Gen Z and its importance for the sustainability of urban destinations. The number of young tourists has recorded one of the fastest growth rates worldwide (UNWTO, 2016; UNWTO & WYSE Travel Confederation, 2010), and among them, backpackers are a growing segment of the tourism industry. If there were no COVID-19 pandemic, numbers would probably have reached 370 million young tourists during 2020, including the backpacker tourist segment, which would have been responsible for expenses in the order of USD 400 billion (UNWTO, 2016).

Data collection was performed using a questionnaire survey and GPS tracking of movements made during a day visit to Porto in Portugal. A total of 74 tourist trip itineraries (Smartphone GPS tracks) and questionnaires from Gen Z and Gen Y backpackers were collected. The 74 tracks obtained in this study correspond to 29.36% of the total participants (n = 252). To the best of our knowledge, this is the first study on Gen Z and Gen Y space–time behaviour using GPS technology to track their movements for a one-day visit to an urban destination. This research reveals some important conclusions that will help support tourism destination managers. The spatial analysis of movements performed by both generations doesn't reveal a significative difference, since both generations of backpackers consume the urban destination in a very similar way. However, Gen Z and Y reveal a different pattern of temporal behaviour: Gen Z backpackers made short-duration visits compared to those of Gen Y, who spent more time visiting the parishes located in the historical centre of the urban destination and classified by UNESCO as World Heritage. Backpackers belonging to Gen Z also reveal a greater territorial amplitude covering longer distances which can be explained by looking for off-the-beaten tracks and seeking out remote places (Robinson & Schänzel, 2019).

DOI: 10.4324/9781003289586-8

This chapter is structured into five sections. After this introduction, the second section discusses the literature review focused on two main themes, Gen Z, its evolution, characteristics and main motivations and tourists' space–time behaviour. The third section is dedicated to describing the methodology and data collection. The fourth section is related to the analysis and discussion of results, and finally, the fifth section presents the main conclusions of the study, contributions, and main limitations.

Literature review

Gen Z: evolution and characteristics

Twenge et al. (2010) refer that people born in the same chronological, social and historical timeframe are collectively called a generation, and because of that, according to the generation theory they share similar characteristics and basic behavioural profiles. Generation-based research identifies different groups of consumers and their specific needs and desires (Chhetri et al., 2014), and researchers always tend to characterise and locate it in a timeframe of different generational groups. Since the Baby Boomers (born between 1946 and 1964), Generation X (born between 1965 and 1980), Millennials or Gen Y (born between 1981 and 1995), and Gen Z or iGen or post-Millennials (born between 1995 and 2012). Over the years new generations have emerged, and similar to the previous generation a number of terms have been used to describe them, such as Gen Z, Gen Zers, post-Millennials (Bassiouni & Hackley, 2014), Tweens, Baby Bloomers (Williams et al., 2010), or iGens (Schneider, 2015). Some researchers define Gen Zers as people who were born after 2000 (Armstrong & Kotler, 2017), whereas others as those who were born in the 1996–2010 period (Monaco, 2018).

In tourism and travel, Gen Z is considered an important group (Barnes, 2018) due to two main factors: first, their influence on family holidays; second, their preference for experiences rather than possessions, increasing their propensity to travel and search for funny experiences. Gen Z is found to be vastly different from previous generations because of the major defining moments of their childhood (e.g., 9/11, 2008 financial crisis and the rise of technology; Khatri & Dixit, 2016; Rodriguez et al., 2019). Furthermore, they are open-minded, bucket-list oriented and look for off-the-beaten-path locations, seeking out remote places and engaging in numerous travels/activities (Robinson & Schänzel, 2019). Gen Z members are budget-conscious travellers and usually start off their travel without a set destination in mind. All these characteristics also meet the literature's descriptions of backpackers, a tourist segment, made up of predominantly young travellers who plan and prepare their own trip, seeking direct cultural contact, novelty, spontaneity and risk-taking around the world (Martins & da Costa, 2022). Their high social and mobile applications penetration rates, large market size, high-income level and readiness for new technologies make them more appropriate for generalising the study findings on a global scale (Dimitriou & AbouElgheit, 2019).

According to Dimitriou and AbouElgheit (2019), a limited number of studies aim to understand Gen Zers' trends and preferences in travel (Turen, 2015; Bradley, 2016; Fuggle, 2017; Trend Watch, 2017). Some studies investigated the impact of specific issues such as social media, word-of-mouth and travel planning on the travel buying decision process (Pinto et al., 2015; Xiang et al., 2015). Although several publications have discussed their general and overall characteristics, few studies have explored their behaviour, preferences and habits as travellers and guests, namely the way they consume the urban destinations they visit, or in other words, their spatiotemporal behaviour.

Since this new generation will impact the hospitality and tourism sector, it is of fundamental importance for hospitality and tourism marketers to examine this emerging market segment, analysing their attitudes and needs, and the major differences between generations of travellers (Turen, 2015), particularly between Gen Y and Gen Z. Demographic trends and changes in values and perceptions have a strong influence on consumer behaviour and marketing strategies (Yeoman et al., 2012).

Gen Z has been raised during changes occasioned by the Internet, smartphones, laptops, freely available network, and digital media. Born into a digital age and with increasing international travel, this young generation is likely to transform tourism and destinations (Robinson & Schänzel, 2019). This chapter is mainly focused on a group of tourists belonging to Gen Z and Y who travel with backpacks on their backs and with a preference for budget accommodation.

Motivations and tourists' space–time behaviour

The motivation to travel can be described as a broad concept, which encompasses biological, psychological, and social factors that activate, guide and maintain the behaviour in different degrees of intensity. The literature review focusing on backpackers' motivations to travel has revealed the existence of a large and diverse set of motivations that reflect the complexity and heterogeneity of this market segment in line with the work of Oliveira-Brochado and Gameiro (2013) who note that there is an emerging diversity and growing heterogeneity in the preferences of backpackers, known as tourists that enjoy different types of experiences during their visits.

Although tourism in urban spaces is one of the most popular forms of tourism, the spatiotemporal behaviour of tourists in urban spaces remains under-researched (Shoval et al., 2011). Understanding the movements made and their relationships with the decisions that tourists make about where, how and at what places and times they move from one attraction to another remains a process of great complexity because there are several factors influencing it (Lew & McKercher, 2006; Xia et al., 2011). As Shoval et al. (2015, p. 80) point out: "Human spatial behavior is the sum of three parallel dimensions: 'what', 'when' and 'where'. The 'what' describes the type of activity performed, the 'when' the temporal dimension of that activity and the 'where' its spatial element".

The existing literature on the spatiotemporal behaviour of tourists has focused on the choices of the destination, the experience of tourists, their segmentation and the way the destination is consumed (Grinberger et al., 2014), on the cultural origin of tourists (Dejbakhsh et al., 2011), on tourists who visit the destination for the first time and repeatedly (Caldeira & Kastenholz, 2018; McKercher et al., 2012), on the first and last day of the visit (Mckercher & Lau, 2008) and on sociodemographic characteristics (Espelt & Benito, 2006; Tchetchik et al., 2009; Xia et al., 2010). Other studies focus on the satisfaction of tourists (Caldeira, 2014), the impact of the distance travelled from the country of origin to the country of destination (Caldeira & Kastenholz, 2015) or on environmental sustainability (Dickinson et al., 2013; Edwards & Griffin, 2013).

To analyse the space–time behaviour of tourists, researchers have identified a diverse set of factors with an impact on the spatiotemporal movement patterns. According to Caldeira (2014) and Caldeira and Kastenholz (2020), tourists' spatiotemporal behaviour in the urban intra-destination context must take into account the characteristics of the tourists, the characteristics of the trip and the characteristics of the visited destination, as they analyse the movement (territoriality, linearity, locomotion and guidance) and multi-attraction (intensity and specificity) of the tourists. Regardless of the destination, the characteristics of tourists – namely their sociodemographic particularities such as age, gender, education, country of origin, level of education and income, among others – are variables that have been frequently used in several studies on spatiotemporal behaviour of tourists (Caldeira, 2014; De Cantis et al., 2016; Edwards et al., 2009; Espelt & Benito, 2006; Hunt & Crompton, 2008; Le-Klähn et al., 2015; Xia et al., 2010; Zakrisson & Zillinger, 2012).

Data and methodology

Data collection was performed using a questionnaire survey and GPS tracking of movements made during a day visit. The combination of the two methods (GPS tracking and questionnaire) has become common in studies on the spatiotemporal behaviour of tourists (Ferrante et al., 2018; Martins et al., 2022; Yun & Park, 2015) as it allows the collection of rich and rigorous information (Zakrisson & Zillinger, 2012). In the morning, the tourists were approached randomly at the reception of the hostels and after a short explanation of the research objectives, the researcher asked for their collaboration, requesting the installation of the Open GPS Tracker app for Android operating systems, and the Simple Logger app for iOS operating systems. Backpackers were asked to start up their apps as soon as they left the accommodations and these GPS Apps automatically recorded the coordinates (latitude, longitude and altitude) of the wearers every 10 seconds, which can reflect their movements throughout their entire tours. All information collected remained anonymous and tracking was not done online or in real time. The researcher would only have access to the data collected after it was sent to their email address. Upon arrival at the accommodation, participants responded to a questionnaire.

The use of smartphones in academic studies on space–time behaviour is still at an early stage (Shoval & Ahas, 2016). With a set of various sensors and technologies, smartphones are a true hybrid tracking device with broad potential but as Thimm and Seepold (2016) found, most tourists approached who agreed to participate in their study did not download the application to their smartphones. However, Yun and Park (2015) successfully used a GPS app to analyse the spatial and temporal movement of visitors to a festival in a rural area in South Korea, with 72.6% of participants sending the information collected to researchers during the five days of the event (66 tracks).

A total of 74 tourist trip itineraries (Smartphone GPS tracks) and questionnaires from Gen Z and Gen Y backpackers were successfully collected. The 74 tracks obtained in this study correspond to 29.36% of the total participants (n = 252), a participation rate well above the 15% recorded by Miyasaka et al. (2018), and Yun and Park (2015), or in the study conducted by Shoval and Isaacson (2007) in the ancient city of Akko, Israel, where only 40 GPS tracks were used (16.26% of the total).

Data were analysed using the IBM SPSS software, version 22, for Windows (IBM Corp, 2013), and the cartography was developed using the free open-source software QGIS 3.12.2. With the georeferenced data obtained by GPS tracking, two thematic maps were built to analyse how the Porto urban area was explored by Gen Z and Gen Y visitors in order to understand their spatial activity. Porto municipality was divided into a raster (a grid 100 × 100 metres) and the number of visitors that passed through each cell in the grid was counted (Shoval et al., 2009). The high-traffic areas are dark and the low-traffic areas are light.

Results and discussion

With a total area of 41 km² and a population of 231,828 (Instituto Nacional de Estatística, 2022), Porto is a Portuguese city classified as a World Heritage Site by UNESCO in 1996 (Figure 6.1). In recent years, the city has become an increasingly popular tourist destination, recording more than 4.5 million overnight stays in 2019, before the COVID-19 pandemic (Pordata, 2022).

A key element of successful tourism is the ability to recognise and deal with change across a wide range of key external factors (economic, political, environmental, social and demographic) and the way they interact. Demographic changes are one of the most important key factors that can affect tourism directly and indirectly (Grimm et al., 2009), and the future growth of the tourism sector will depend to some extent on how the industry understands the social and demographic trends and how they will influence tourist behaviour.

Table 6.1 presents a comparison between the most important sociodemographic characteristics of Gen Y and Z participants. The sample consisted of 24 (32.4%) Gen Z tourists and 50 (67.6%) Gen Y (58% female and 42% male participants), mostly from European countries. Both generations of tourists were travelling alone and the main differences in the travel group can be found in Gen Z, where 33% of individuals were travelling with a close friend. Gen Y tourists express a preference for travelling with a girlfriend/boyfriend (18%).

The analysis of Table 6.1 also allows the researchers to conclude that the most significant difference exists in the average length of stay with Gen Y backpackers spending more nights in a destination (7.4) which can be explained by the fact that

Figure 6.1 Porto municipality and Santa Marinha parish (Vila Nova de Gaia municipality)

Table 6.1 Sociodemographic characteristics

Indicators	Gen Y		Gen Z	
	N.º	%	N.º	%
Gender				
Male	26	52.0%	13	54.2%
Female	24	48.0%	11	45.8%
World tourism regions				
Europe	37	74.0%	15	62.5%
America	10	20.0%	6	25.0%
Eastern Asia	3	6.0%	3	12.5%
Annual income				
No annual income (€)	0	0.0%	8	33.3%
< 3000	5	10.0%	5	20.8%
3001–7000	3	6.0%	5	20.8%
7001–14000	6	12.0%	0	0.0%
14001–20000	4	8.0%	4	16.7%
20001–30000	12	24.0%	2	8.3%
> 30000	14	28.0%	0	0.0%
Travel group				
Travelling alone	28	56.0%	13	54.2%
Travelling with a close friend	6	12.0%	8	33.3%
Travelling with a group of close friends	5	10.0%	2	8.3%
Travelling with girlfriend/boyfriend	9	18.0%	1	4.2%
Travelling with relatives	2	4.0%	0	0.0%
Trip typology				
Extended holidays from studies/work	17	34.0%	3	12.5%
Gap year after graduating	1	2.0%	1	4.2%
Gap year before graduating	0	0.0%	5	20.8%
Short holiday from studies/work	23	46.0%	6	25.0%
Short package trips	5	10.0%	4	16.7%
Study abroad	2	4.0%	3	12.5%
Volunteering	2	4.0%	1	4.2%
Other	0	0.0%	1	4.2%
Average length of stay (No. of nights)		7.4		5.8

Source: Own construction.

they have higher incomes than Gen Z (52% have an annual income superior to 20,000€). It is also important to note that 75% of Gen Z individuals have an annual income inferior to 7,000€ (33% with no annual income; 20.8%, inferior to 3,000€ and 20.8% between 3,001€ and 7,000€, respectively). The trip typology also indicates some significant differences between the two generations with 80% of Gen Y participants mentioning that they are on short or extended holidays from studies/work, compared to 37.5% of Gen Z. Around 21% of Gen Z mentioned that were in a gap year before graduating and 16.7% in a short package trip.

Crossing the tourists tracked movements with the main attractions of the city (Figure 6.1), there is a higher movement intensity in Ponte D. Luís, na Sé Catedral, Casa do Infante, Palácio da Bolsa, Mercado Ferreira Borges, Estação de São Bento,

Mercado do Bolhão, e Jardins do Palácio de Cristal. Porto is a city known for its rich heritage and as a consequence visits are based on visiting monuments, as demonstrated by Espelt and Benito (2018).

It is clear in the maps of Figure 6.2 that the main concentration of tourists is Liberdade/Aliados square where the hostel *Nice Way* is located, an accommodation of a large number of participants of this research. As they move away from the parishes where the historical centre is located, the density of the movements strongly decreases, with the parishes located on the northern edge of the municipality having a reduced number of visits, or even none at all, as seen in Aldoar. As happened in Lisbon, there is also in Porto a coincidence in the distribution of the tracked movements "that becomes more open as we move away from the city centre to its periphery and from this to the metropolitan area" (Caldeira, 2014, p. 267).

Besides the number of tourist attractions not being very relevant in areas further away from the historical centre of Porto, the public transport network, especially the Metro, does not guarantee the same accessibility to the whole city. As there is an interrelation between the mode of transport and the spatial extent of tourists' visits (Le-Klähn et al., 2015) it is not difficult to explain a higher intensity of visits in areas served by metro stations. The city streets were thus used for circulation without significant stops or experiences (Yun et al., 2018).

Through the analysis of the graphic in Figure 6.2, it is possible to verify that the backpackers belonging to Gen Z spent more time visiting the parishes located in the historical centre (Santo Ildefonso, São Nicolau and Sé) classified by UNESCO as World Heritage. Comparing the location of the main attractions in the city of Porto (Figure 6.1) with the map of the intensity of movements (Figure 6.2) it can be concluded that backpackers from Gen Z express a preference for visiting the main historical monuments of the city.

In this research, participants belonging to Gen Y have made long-duration visits (304.1 min.) compared to Gen Z backpackers (247.7 min). However, the average distance travelled was higher among Gen Z participants which can be explained by the visits made by two tourists to attractions located at very distant places such as the well-known Região Demarcada do Douro, the oldest demarcated and regulated wine region in the world.

As shown in Table 6.2, backpackers belonging to Gen Z reveal a greater territorial amplitude covering longer distances as can be observed by the average distance travelled (18.2 km) which can be explained by looking for off-the-beaten tracks, and seeking out remote places (Robinson & Schänzel, 2019). Covering longer distances implies using more public transportation and/or rented cars that result in a higher average speed, a higher altitude variation and a higher accumulates climb (Table 6.2). It is also important to note that some of the Gen Z backpackers visited with greater incidence other places located outside the Porto, namely, Vila Nova de Gaia, Matosinhos, Guimarães, Aveiro, and the Região Demarcada do Douro.

According to the questionnaire administered in this study, the most popular activities amongst Gen Y backpackers were walking around (84.5%), eating local food at a restaurant/café (44,5%), visiting attractions (e.g., monuments, museums, exhibitions) with 40.5% of respondents selecting this option, and shopping

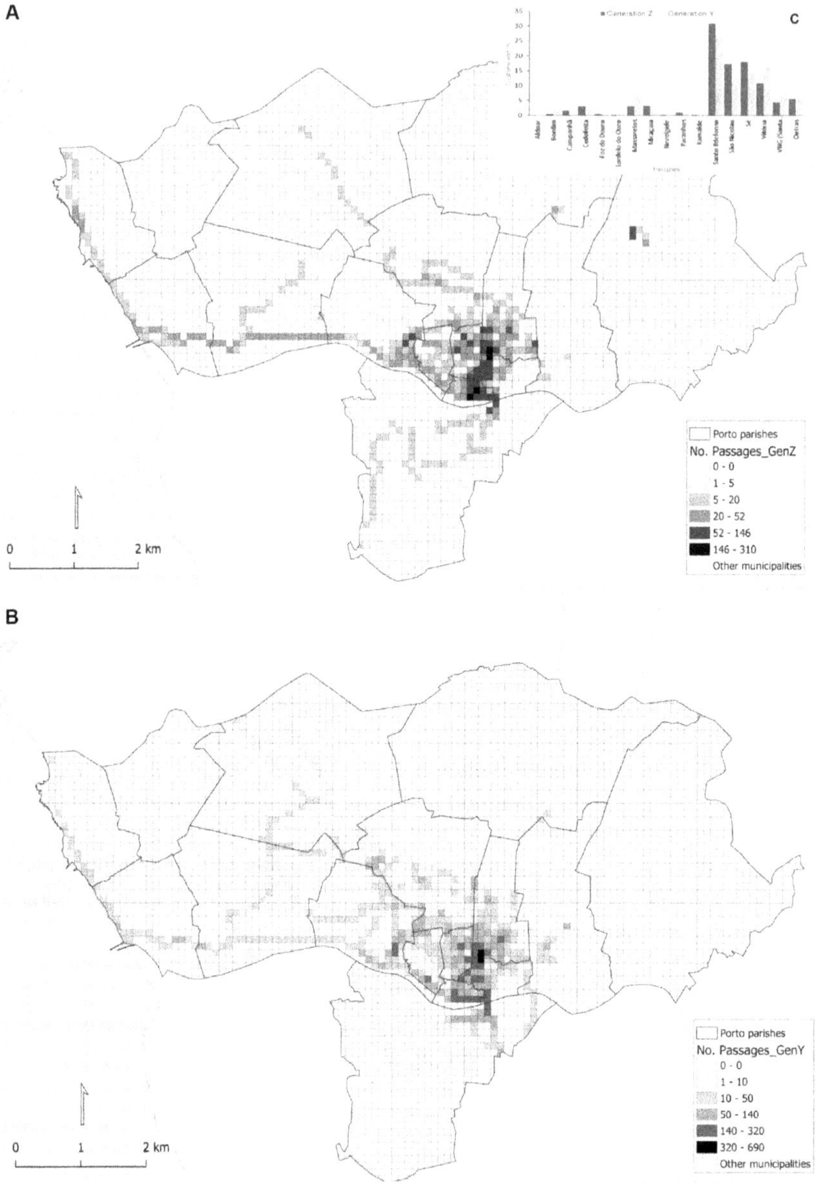

Figure 6.2 Movement intensity of Gen Z backpackers (A) and Gen Y backpackers (B) in terms of number of passages; total visit time (%) spent in each parish is presented in (C)

Table 6.2 Gen Z and Gen Y spatiotemporal behaviour

		Day visit duration (min)	Time in motion (min)	Time in motion (%)	Travelled distance (day visit; km)	Maximum dispersal from accommo- dation (km)	Average speed (km/h)	Altitude variation (m)	Accumu- lated climb (m)
Gen Z	M	**247.7**	**127.0**	**55.0**	**18.2**	**7.0**	**3.0**	**101.1**	**432.3**
	Md	233.5	93.0	49.0	5.1	1.2	1.8	77.0	140.5
	SD	162.6	92.1	23.9	51.9	24.4	3.9	146.9	1230.2
Gen Y	M	**304.1**	**124.3**	**48.8**	**11.0**	**3.7**	**2.5**	**81.4**	**212.1**
	Md	283.7	94.6	43.0	5.4	1.5	1.9	89.0	142.0
	SD	224.8	112.4	24.6	18.2	10.5	2.0	31.3	237.7

Note: M – Mean; Md – Median; SD – Standard deviation.

Source: Own construction.

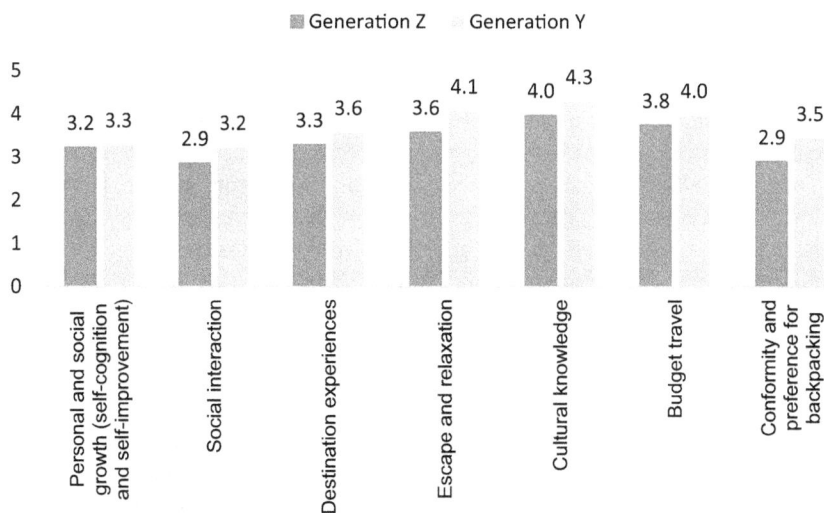

Figure 6.3 Motivations to visit Porto
Source: Own construction.

(16.3%). The most popular activities amongst Gen Z backpackers were walking around (83.6%), visiting attractions (53.6%), practising sports (e.g., surfing, canoeing), and going to the beach (10.5%).

An analysis of the motivations for visiting the urban destination of Porto (Figure 6.3) shows some differences between the two generations. When asked about the level of agreement with the motivations to visit Porto on a scale of 1 to 5 (1= not important at all; 5= very important), Gen Y backpackers reveal a higher degree of agreement in all motivations.

The most important motivations for Gen Z backpackers are cultural knowledge and budget travel. Knowing and understanding the local culture, history and society and exploring other cultures and improving knowledge are very important to Gen Z individuals when visiting Porto, a World Heritage city classified by UNESCO. They are also known as budget-conscious travellers (Southan, 2017) and not surprisingly, Gen Z backpackers visited Porto because it is a value-for-money destination, allowing them to travel on a budget. For Gen Y visitors, cultural knowledge was very important too, followed by escape and relaxation motivation that could be related to the fact that most of its members are older than those of Gen Z and are more established in relation to their professional careers.

Conclusions

As far as it was possible to identify, this is the first study on Gen Z and Gen Y space–time behaviour using GPS technology to track their movements during a one-day visit to an urban destination. In this research, the main attractions visited by backpackers were: the bridge D. Luís, Sé Catedral, Casa do Infante, Palácio da Bolsa, Mercado Ferreira Borges and the train station of São Bento. All these attractions are located in the historic centre. The old and traditional Mercado do Bolhão and the gardens of Palácio de Cristal were also important attractions for both generations. The spatial analysis of movements performed by both generations doesn't reveal a significant difference. Both generations of backpackers consume the urban destination in a very similar way. As visitors move away from the parishes where the historical centre is located, the density of the movements strongly decreases, with the parishes located on the northern edge of the municipality having a reduced number of visits, or even none at all. However, Gen Z and Y reveal a different pattern of temporal behaviour: Gen Z backpackers made short-duration visits compared to those of Gen Y, spending more time visiting the parishes located in the historical centre (Santo Ildefonso, São Nicolau and Sé) classified by UNESCO as World Heritage.

Backpackers belonging to Gen Z also reveal a greater territorial amplitude covering longer distances which can be explained by looking for off-the-beaten tracks and seeking out remote places (Robinson & Schänzel, 2019). Covering longer distances implies using more public transportation and/or rented cars that result in a higher average speed, a higher altitude variation and a higher accumulates climb.

The most popular activities amongst Gen Y and Z backpackers were walking around and visiting attractions. Gen Z backpackers also reveal a lower degree of agreement in all motivations and the most important motivations to Gen Z backpackers are cultural knowledge and budget travel.

The main limitation of this study is the sample size. Only 74 tourist trip itineraries (Smartphone GPS tracks) and questionnaires from Gen Z and Gen Y backpackers were successfully collected (29.36% of the total participants), a number above the 15% recorded by Miyasaka et al. (2018), and Yun and Park (2015) or in the study conducted by Shoval and Isaacson (2007) in the ancient city of Akko, Israel. It is also important to highlight that only one-day of the visit was tracked and not the entire stay of the participants, in line with the main studies that have been carried out internationally on the spatiotemporal behaviour of tourists.

References

Armstrong, G., & Kotler, P. (Eds.). (2017). *Marketing: An introduction.* Pearson Education.
Barnes, R. (2018). Gen-Z expert panel the "little extraordinaires" to consult for Royal Caribbean. *Cruise Trade News.* www.cruisetradenews.com/gen-z-expert-panel-the-little-extraordinaires-to-consult-for-royal-caribbean/
Bassiouni, D. H., & Hackley, C. (2014). Generation Z children's adaptation to digital consumer culture: A critical literature review. *Journal of Customer Behaviour, 13*(2), 113–133.
Bradley, D. (2016). The new influencers. *Haymarket Media.* www.prweek.com/article/1379310/new-influencers
Caldeira, A. M. (2014). *A experiência de visita dirigida a múltiplas atrações: Análise do comportamento espacial do turista e da sua satisfação.* http://ria.ua.pt/handle/10773/12755
Caldeira, A. M., & Kastenholz, E. (2015). Spatiotemporal behaviour of the urban multi-attraction tourist: Does distance travelled from country of origin make a difference? *Tourism & Management Studies, 11*(1), 91–97.
Caldeira, A. M., & Kastenholz, E. (2018). Tourists' spatial behaviour in urban destinations: The effect of prior destination experience. *Journal of Vacation Marketing, 24*(3), 247–260.
Caldeira, A. M., & Kastenholz, E. (2020). Spatiotemporal tourist behaviour in urban destinations: A framework of analysis. *Tourism Geographies, 22*(1), 22–50.
Chhetri, P., Hossain, M. I., & Broom, A. (2014). Examining the generational differences in consumption patterns in Southeast Queensland. *City, Culture and Society, 5*(4), 1–9.
De Cantis, S., Ferrante, M., Kahani, A., & Shoval, N. (2016). Cruise passengers' behavior at the destination: Investigation using GPS technology. *Tourism Management, 52*, 133–150.
Dejbakhsh, S., Arrowsmith, C., & Jackson, M. (2011). Cultural influence on spatial behaviour. *Tourism Geographies, 13*(1), 91–111.
Dickinson, J. E., Filimonau, V., Cherrett, T., Davies, N., Norgate, S., Speed, C., & Winstanley, C. (2013). Understanding temporal rhythms and travel behaviour at destinations: Potential ways to achieve more sustainable travel. *Journal of Sustainable Tourism, 21*(7), 1070–1090.
Dimitriou, C. K., & AbouElgheit, E. (2019). Understanding Generation Z's travel social decision-making. *Tourism and Hospitality Management, 25*(2), 311–334.
Edwards, D., & Griffin, T. (2013). Understanding tourists' spatial behaviour: GPS tracking as an aid to sustainable destination management. *Journal of Sustainable Tourism, 21*(4), 580–595.
Edwards, D., Griffin, T., Hayllar, B., & Dickson, T. (2009). *Using GPS to track tourists spatial behaviour in urban destinations.* http://papers.ssrn.com/sol3/papers.cfm?abstract_id=1905286
Espelt, N. G., & Benito, J. A. D. (2006). Visitors' behavior in heritage cities: The case of Girona. *Journal of Travel Research, 44*(4), 442–448.
Espelt, N. G., & Benito, J. A. D. (2018). First-time versus repeat visitors' behavior patterns: A GPS analysis. *Boletín de la Asociación de Geógrafos Españoles, 78*, 49–65.
Ferrante, M., De Cantis, S., & Shoval, N. (2018). A general framework for collecting and analysing the tracking data of cruise passengers at the destination. *Current Issues in Tourism, 21*(12), 1426–1451.
Fuggle, L. (2017). Gen Z will soon transform the travel industry. *Huffington Post.* www.huffingtonpost.com/lucy-fuggle/gen-z-are-going-to-transf_b_9870028.html
Grimm, B., Metzler, D., Butzmann, E., & Schmücker, D. J. (2009). The impact of demographic change on tourism demand structures in Germany and selected source markets.

The future travel volume and behavior of different age groups. *TW Zeitschrift für Touris-muswissenschaft, 2*(2), 111–132.

Grinberger, A. Y., Shoval, N., & McKercher, B. (2014). Typologies of tourists' time – space consumption: A new approach using GPS data and GIS tools. *Tourism Geographies, 16*(1), 105–123.

Hunt, M. A., & Crompton, J. L. (2008). Investigating attraction compatibility in an East Texas City. *International Journal of Tourism Research, 10*(3), 237–246.

IBM Corp. (2013). *IBM SPSS statistics for windows, version 22.0.* IBM Corp.

Instituto Nacional de Estatística. (2022). *Censos 2021.* Resultados Provisórios. www.ine.pt/scripts/db_censos_2021.html

Khatri, K., & Dixit, N. (2016). Managing aspiration of Generation "Y" and Generation "Z" at work place. *Khoj Journal of Indian Management Research & Practices, 5,* 65–69.

Le-Klähn, D. T., Roosen, J., Gerike, R., & Hall, C. M. (2015). Factors affecting tourists' public transport use and areas visited at destinations. *Tourism Geographies, 17*(5), 738–757.

Lew, A., & McKercher, B. (2006). Modeling tourist movements: A local destination analysis. *Annals of Tourism Research, 33*(2), 403–423.

Martins, M. R., & da Costa, R. A. (Eds.). (2022). *The backpacker tourist: A contemporary perspective.* Emerald Group Publishing.

Martins, M. R., da Costa, R. A., & Moreira, A. C. (2022). Backpackers' space – time behavior in an urban destination: The impact of travel information sources. *International Journal of Tourism Research, 24*(3), 456–471.

Mckercher, B., & Lau, G. (2008). Movement patterns of tourists within a destination. *Tourism Geographies, 10*(3), 355–374.

McKercher, B., Shoval, N., Ng, E., & Birenboim, A. (2012). First and repeat visitor behaviour: GPS tracking and GIS analysis in Hong Kong. *Tourism Geographies, 14*(1), 147–161.

Miyasaka, T., Oba, A., Akasaka, M., & Tsuchiya, T. (2018). Sampling limitations in using tourists' mobile phones for GPS-based visitor monitoring. *Journal of Leisure Research, 49*(3–5), 298–310.

Monaco, S. (2018). Tourism and the new generations: Emerging trends and social implications in Italy. *Journal of Tourism Futures, 4*(1), 7–15.

Oliveira-Brochado, A., & Gameiro, C. (2013). Toward a better understanding of backpackers' motivations. *Tékhne, 11*(2), 92–99.

Pinto, N. P., Gomes, D. D. O., Cavalcante, F., Mendes, G. A., & Sales, R. K. L. (2015). The influence of social networks on consumers' buying decision process-a study of tourism products. *Asian Journal of Business and Management Sciences, 4*(3), 1–10.

Pordata. (2022). *Alojamentos turísticos: Dormidas por tipo de alojamento. Ocupação de Alojamentos Turísticos.* www.pordata.pt/Subtema/Municipios/Ocupação+de+Alojamentos+Turísticos-361

Robinson, V. M., & Schänzel, H. A. (2019). A tourism inflex: Generation Z travel experiences. *Journal of Tourism Futures, 5*(2), 127–141.

Rodriguez, M., Boyer, S., Fleming, D., & Cohen, S. (2019). Managing the next generation of sales, Gen Z/millennial cusp: An exploration of grit, entrepreneurship, and loyalty. *Journal of Business-to-Business Marketing, 26*(1), 43–55.

Schneider, J. (2015). How to market to the iGeneration. *Harvard Business Review.* https://hbr.org/2015/05/how-to-market-to-the-igeneration?utm_source=feedburner&utm_medium=feed&utm_campaign=Feed

Shoval, N., & Ahas, R. (2016). The use of tracking technologies in tourism research: The first decade. *Tourism Geographies, 18*(5), 587–606.

Shoval, N., & Isaacson, M. (2007). Sequence alignment as a method for human activity analysis in space and time. *Annals of the Association of American Geographers, 97*(2), 282–297.

Shoval, N., Isaacson, M., & Birenboim, A. (2009, November 3). Monitoring impacts of visitors with aggregative GPS data. In B. Gottfri, N. V. Weg, R. Billenand, & P. De Maeyer (Eds.), *Proceedings of the 3rd workshop on behaviour monitoring and interpretation (BMI'09), Ghent, Belgium* (pp. 33–46). BMI.

Shoval, N., McKercher, B., Birenboim, A., & Ng, E. (2015). The application of a sequence alignment method to the creation of typologies of tourist activity in time and space. *Environment and Planning B: Planning and Design, 42*(1), 76–94.

Shoval, N., McKercher, B., Ng, E., & Birenboim, A. (2011). Hotel location and tourist activity in cities. *Annals of Tourism Research, 38*(4), 1594–1612.

Southan, J. (2017). From boomers to Gen Z: Travel trends across the generations. *Globetrender Magazine*. http://globetrendermagazine.com/2017/05/19/travel-trends-across-generations/

Tchetchik, A., Fleischer, A., & Shoval, N. (2009). Segmentation of visitors to a heritage site using high-resolution time-space data. *Journal of Travel Research, 48*(2), 216–229.

Thimm, T., & Seepold, R. (2016). Past, present and future of tourist tracking. *Journal of Tourism Futures, 2*(1), 43–55.

Trend Watch. (2017). *Generation Z: The new destination deciders*. www.travelagentmagazinedigital.com/publication/frame.php?i=403417&p=&pn=&ver=htm 15&view=article Browser&article_id=2769308

Turen, R. B. (2015). It ain't all about the millennials. *Travel Weekly*. www.travelweekly.com/Richard-Turen/It-aint-all-about-the-millennials

Twenge, J. M., Campbell, S. M., Hoffman, B. J., & Lance, C. E. (2010). Generational differences in work values: Leisure and extrinsic values increasing, social and intrinsic values decreasing. *Journal of Management, 36*(5), 1117–1142.

UNWTO. (2016). *Global report on the power of youth travel: Affiliate members report* (Vol. 13). www.wysetc.org/wp-content/uploads/2016/03/Global-Report_Power-of-Youth-Travel_2016.pdf

UNWTO & WYSE Travel Confederation. (2010). *The power of youth travel: AM reports* (Vol. 2). www.wysetc.org/wp-content/uploads/2014/12/wysetc-unwto-report-english_the-power-of-youth.pdf

Williams, K. C., Page, R. A., Petrosky, A. R., & Hernandez, E. H. (2010). Multi-generational marketing: Descriptions, characteristics, lifestyles, and attitudes. *The Journal of Applied Business and Economics, 11*(2), 21.

Xia, J. C., Evans, F. H., Spilsbury, K., Ciesielski, V., Arrowsmith, C., & Wright, G. (2010). Market segments based on the dominant movement patterns of tourists. *Tourism Management, 31*(4), 464–469.

Xia, J. C., Zeephongsekul, P., & Packer, D. (2011). Spatial and temporal modelling of tourist movements using Semi-Markov processes. *Tourism Management, 32*(4), 844–851.

Xiang, Z., Magnini, V. P., & Fesenmaier, D. R. (2015). Information technology and consumer behavior in travel and tourism: Insights from travel planning using the internet. *Journal of Retailing and Consumer Services, 22*, 244–249.

Yeoman, I. (2012). *2050-tomorrow's tourism* (Vol. 55). Channel View Publications.

Yun, H. J., Kang, D. J., & Lee, M. J. (2018). Spatiotemporal distribution of urban walking tourists by season using GPS-based smartphone application. *Asia Pacific Journal of Tourism Research, 23*(11), 1047–1061.

Yun, H. J., & Park, M. H. (2015). Time – space movement of festival visitors in rural areas using a smart phone application. *Asia Pacific Journal of Tourism Research*, *20*(11), 1246–1265.

Zakrisson, I., & Zillinger, M. (2012). Emotions in motion: Tourist experiences in time and space. *Current Issues in Tourism*, *15*(6), 505–523.

7 The generational transition of Gen Z tourists' behaviour

A sociological snapshot from the Vesuvius National Park

Salvatore Monaco

Introduction

According to the International Union for Conservation of Nature (IUCN) World Commission on Protected Areas (WPCA; 2018), a protected area could be defined as a clearly specified geographical space, recognised, dedicated and managed to achieve the long-term conservation of nature with associated ecosystem services and cultural values. Under this umbrella term, there are national parks, wilderness areas, community conserved areas, and nature reserves of importance for flora, fauna, or their features of geological or other special interests, which are reserved and managed for conservation purposes (Locke & Dearden, 2005; Worboys et al., 2015). These areas may be designated by government institutions in some countries, or by private landowners, such as charities and research institutions in other territories. Protected areas are not only the core of biodiversity conservation, but they also contribute to people's livelihoods, particularly at the local level, since they are able to promote local services and the environment (e.g., Andrade & Rhodes, 2012; Potts et al., 2014; Wells & McShane, 2004). Their role in helping mitigate and adapt to climate change is also increasingly recognised. In the current time of ecological transition, it has been recognised that the global network of protected areas occupies a central role in climate-change mitigation (e.g., Dudley et al., 2010; Elsen et al., 2020; Melillo et al., 2016).

In recent years, many natural protected areas around the world have been committed to promoting tourism sustainability (e.g., Buongiorno & Intini, 2021; Dedeke, 2017; Stone & Nyaupane, 2016). In fact, although combining tourism promotion with the preservation of the natural habitat may seem quite paradoxical, several natural protected areas have been working to create a harmonious relationship between tourist flows and nature through a series of policies, communication activities, public events, and other initiatives (e.g., Bushell & Bricker, 2017; Laven et al., 2015).

In this scenario, the European Charter for Sustainable Tourism in Protected Areas (Europarc Federation, 2010) has been engaged in setting up a model for the management of sustainable tourism, in line with international guidelines on biodiversity and tourism development. Among European countries in this endeavour, Italy certainly represents one of the most virtuous cases, with about 40 protected

DOI: 10.4324/9781003289586-9

areas that have been certified by Europarc as sustainable tourist destinations. From a naturalistic point of view, Italy has a great competitive advantage as it enjoys rich biodiversity given by a perfect combination of sea and mountains and the harmony between man and nature that make possible the development of sustainable tourism.

The current study

Among different Italian regions, Campania stands out most because it hosts an immense protected natural heritage, made up of 123 Sites of the Natura 2000 Network, which was founded by the European Union in 1992 for the protection and conservation of habitats and animal and plant species (Maiorano et al., 2007). More specifically, Campania Region has 12 Regional Parks, Nature Reserves, and two National Parks. The first is the Cilento and Vallo di Diano National Park, which extends from the Tyrrhenian coast to the foot of the Campania-Lucanian Apennines. Such an extension makes it the second-largest park in Italy. The other National Park in the region is the Vesuvius National Park, southeast of Naples. It was established in 1995 around the Somma-Vesuvius volcanic complex for the purpose of preserving animal and plant species, geological singularities, palaeontological formations, biological communities and biotopes, saving scenic and panoramic values, and striking the hydraulic, hydrogeological, and ecological balances in the Vesuvian territory. In addition to being of great landscape and geological interest, the Park was also born to safeguard Vesuvius, one of the most famous volcanoes in the world. These two parks together with the area that includes Pompeii and the Vesuvian villas are part of the Man and the Biosphere (MAB) programme, launched by UNESCO at the beginning of 1971 with the aim of protecting and promoting the relationship between man and the environment, in compliance with the practices of sustainable development focused on the protection of biodiversity (Dyer & Holland, 1988). The five National Nature Reserves are present in the province of Caserta and in the metropolitan city of Naples and Salerno. In these areas, there are also five Marine Protected Areas. Between the municipalities of Bacoli and Pozzuoli is the Submerged Archaeological Park of Baia, which extends along the coasts of the Gulf of Naples (Paoletti et al., 2005).

The present research work concentrates on the Vesuvius National Park. Given the potential role of young tourists in contributing to the development of sustainable tourism, this paper focuses on the answers collected among the tourists belonging to Generation Z (Gen Z, also known as the Net Generation or the iGeneration) who decided to visit this protected area between 2020 and 2021.

The group of people born roughly from the end of the 1990s through to the early 2010s represent an interesting target for investigation, since they seem more aware of and concerned about the depletion of natural resources, unlike previous generations who were focused on economic growth (Dabija et al., 2020; Lazányi & Bilan, 2017; Yamane & Kaneko, 2021; Williams et al., 2010). They are now entering the workplace and represent the workforce of the future. Currently, Gen Z constitutes one of the fastest-growing segments of international tourism.

Against this backdrop, this paper aims to give an empirical understanding of Gen Zers' profiles, studying on one hand if they are socially and environmentally conscious and, on the other hand, evaluating how much their consumption behaviour and tourist choices could be considered sustainable and oriented towards "green living". In other words, the research aims to understand if, in general, the young tourists participating in the study are attentive to the issue of sustainability in their everyday life, and if so, in what form how they have practised sustainable tourism in the Vesuvius National Park.

The research was conducted by a multidisciplinary research group of OUT (research centre on tourism of the University of Naples Federico II) along the open paths of the Vesuvius National Park at the time of the study (1. "The Valley of Hell"; 2. "Along the Cognoli"; 5. "The Great Cono"; 7. "The Profica Valley"; 9. "River of Lava"). Between 2001 and 2003 the Park established a network of 11 paths that form a total of a 54-km walk, being part of the project "The Paths of the Vesuvius National Park" (see www.parconazionaledelvesuvio.it). Due to the outbreak of a forest fire in 2017, some of the trails were closed or are currently under maintenance.

These paths differ in structure and function. In particular, there are six natural circular trails (n. 1, 2, 3, 4, 5, and 8), one educational path (n. 9), one sightseeing path (n. 6) and one agricultural path (n. 7). The most popular path is n. 5 ("The Great Cono") since it is the only one that gives visitors access to the Crater's lip. In 2019 it received 756,572 paying visitors, an increase of 14% compared to 2018, equal to 665,945 admissions (Corbisiero & Monaco, 2021). Those who visit for free, however, must also be added to these presences.

Methods

To answer the research questions, the working group prepared a structured questionnaire with the aim of collecting data and information in a standardised way. The questionnaire consists of four sections, each of which has been dedicated to a specific dimension of analysis. More specifically:

- *Section 1.* "Visit to the Vesuvius National Park": The first section consisted of seven questions and was prepared to collect information regarding the behaviour and expectations of tourists who visited the Vesuvius National Park area, investigating different aspects, both general and particular, with respect to the way in which they perceived the experience of the trip within the area.
- *Section 2.* "Visit to the paths": The second section consisted of seven questions dedicated to the visit to the Vesuvius paths. More specifically, this set of questions tended to investigate how the travellers collected information about the paths and activities they could or have embarked on, with whom they shared the experience and if they were aware of the overall path network.
- *Section 3.* "Evaluation of the experience": With the third section, the tourists were stimulated through a series of questions to show an assessment of satisfaction in various dimensions (there were five questions for each area investigated):

health security measures; reception and communication; environmental sustainability; risk and safety; mobility and accessibility.
* *Section 4.* "Socio-demographic information and sustainable behaviours": The section that concluded the questionnaire consisted of nine questions for investigating the socio-personal aspects of tourists.

The questionnaire was validated through a pilot survey involving a sample of 50 randomly selected tourists. The aim of this step was to test the reliability of the questionnaire. It was found by using Cronbach's (1951) Alpha. The reliability test checked whether the respondents' scores on each attribute were related to their scores on the other attributes (Bryman & Bell, 2007).

The reliability coefficients were greater than the threshold value of 0.7 (Nunnally, 1978) for all dimensions investigated, ranging from 0.80 to 0.92. On the basis of feedback, some changes in the questionnaire were also accommodated to facilitate the understanding of some items.

The final version of the questionnaires was distributed in person, along the trail network, and online. Indeed, the questionnaire was also digitised and disseminated on the main social channels of OUT and the Vesuvius National Park. A total of 200 questionnaires were completed by members of Gen Z.

The respondent sampling was non-probabilistic, accidental in nature for various reasons: first, since there was no list of subjects to be involved in the study, it would still have been impossible to proceed with a probabilistic sampling; second, the realisation of personal surveys has certainly been conditioned by the health emergency caused by the pandemic. Many subjects preferred not to participate in the study, or they were willing to do so later, filling in the questionnaire independently via CAWI, to avoid personal contact with unknown people as much as possible.

As amply shown in the literature (Etikan & Bala, 2017), accidental sampling could be an excellent basis for exploratory investigations that intend to photograph emerging phenomena.

Findings

In order to reconstruct the social profile of visitors to the trail network of the Vesuvius National Park, some structural information was collected. The Generation Z samples in this study were aged just over 23 with a gender ratio of 1.22 (55% men vs. 45% women).

In terms of the individual path, it is possible to notice path no. 5 ("River of Lava") attracted younger tourists (average age 18 years) than the other ones. Conversely, path no. 2 ("Along the Cognoli") was frequented by visitors with an average age of 26. Most tourists declared they were still in education or training (77% said they were students).

As for the origin of the tourists, path no. 5 attracted, in the post-pandemic period, the highest percentage of foreign visitors (35.2% of the total), followed by path no. 9. There, 22.2% of visitors declared they came from non-European countries. On the contrary, the other trails mainly attracted Italian tourists who resided in the neighbouring areas of Campania. In terms of previous travel experiences, 54.2% of respondents said they had never visited other volcanic sites (both in Italy and in other countries of the world). By contrast, 65% said they had visited other natural parks and protected natural areas around the world, even before the pandemic.

As a result, it is possible to argue that these Gen Zers chose the Vesuvius National Park because they were driven by the desire to travel while getting close to nature, which is in line with other international surveys (e.g., Çaliskan, 2021; Entina et al., 2021; Fermani et al., 2020; Jiang & Hong, 2021). These studies underline the tendency of the new generations to orient their consumption and tourist choices towards green destinations, where their experiences take shape in contact with the surrounding natural environment.

A specific section of the questionnaire served to uncover the reasons that drove tourists to visit different paths. The analysis of the data highlighted that although belonging to the same path network, each path was able to satisfy different tourist needs. More specifically, "The Valley of Hell" was mainly taken by those who wished to take part in educational activities, such as guided tours and moments of environmental education. This finding is explained by the fact that some local associations often organise excursions along this path in order to instil into visitors' historical information and naturalistic details regarding the rich biodiversity that characterises the entire area. "River of Lava" was mainly frequented by people who declared that they wanted to live a tourist experience in the open air, staying close to the surrounding nature. Unlike the "The Valley of Hell", crowded mainly in the early hours of the day, visitors came to the "River of Lava" path in succession at certain hours: it was frequented by tourists during the day to admire the view, in the afternoon to enjoy the sunset, and in the evening, to take part in cultural events and initiatives organised by associations and guides, or to have an aperitif with friends.

As for path no. 2, almost all visitors said they visited it since they intended to work out while touring the Park. In fact, 30% of respondents said they arrived "Along the Cognoli" riding their bicycle. Other motor and sports activities included trekking and horseback riding. Finally, as regards path number 5, some tourists declared to visit "the Great Cono" out of curiosity as it is at the top of Vesuvius; others, however, wanted to enjoy the breath-taking view of the Gulf of Naples from such a height.

The last section of the questionnaire functioned to evaluate the tourist experience, using a series of parameters. For each element to be evaluated, responses were asked to indicate their level of satisfaction on a 5-point Likert scale (from 1: "not at all satisfied", to 5: "completely satisfied").

From Table 7.1, which shows the average values recorded, it is possible to note that the most appreciated aspects (with a score from 3 upwards) of paths 1, 5, and 9 are "Conservation of biodiversity", "Cleanliness and beautification" and "Trail Maintenance". On the other hand, for all paths (except path no. 5), "Public transportation" recorded the lowest evaluations, with an average score of less than 2. Other negative evaluations were provided by the participants in relation to the lack of bins for separate waste collection and the lack of police officers.

Table 7.1 Customer satisfaction in grades (1–5)

	Path				
	1	*2*	*5*	*7*	*9*
Presence of display boards and indications	2.2	1.7	2.9	2.6	3.5
Cleanliness and beautification of the path	3	2.6	3.4	1.5	3.5
Conservation of biodiversity	3.3	2.9	3.5	2.5	3.7
Trail maintenance	3	1.9	3.4	1.8	3.7
Presence of waste sorting device	2.2	1.6	2.4	1.6	2.2
Headquarters of law enforcement agencies	2.1	2	2.7	1.6	1.8
Barriers and restricted areas	2.5	1.9	3.1	2	2.8
Warning of ruined or dangerous areas	2.3	2.1	3.1	1.7	2.8
Presence of fire alarms or disaster alerts	2.2	1.6	2.8	1.5	2.4
Convenience of public transportation	1.8	1.3	2.6	1.5	1.9
Capacity of parking lots	2.6	2	2.8	2.4	3.2

In general, path no. 9 is indicated as the most appreciated also because it is considered a free alternative to the Great Cono (that is the only route for which the payment of a ticket is required). The path no. 7 is the one that recorded the most negative evaluations and was criticised above all for the poor cleanliness, lack of police control, and lack of path alerts of fire or other disasters.

A further reflection concerns the channels of communication. As reflected in the literature (Koulopoulos & Keldsen, 2016; Monaco, 2021; Turner, 2015; Yussof et al., 2018), young Gen Z are comfortable with digital technology and social media. Having been Internet users since childhood, they show great familiarity with messaging apps on mobile devices or online platforms. The data collected on tourists in the Vesuvius National Park are in line with these assumptions. The results highlight that tourists rely heavily on social networks as their main source of information. The extensive use of online information shows that it is a group of prepared tourists who are fully aware of the characteristics and potential of the place they want to visit. In other words, it is safe to argue that their green attitude is strongly supported by their frequent use of new technologies before, during, and after their travel experiences. In fact, young people involved in the study said they paid a lot of attention to the information they found online.

Although the Vesuvius National Park has its own well organised and resourceful website from which tourists and citizens can find information, tourist comments by word-of-mouth were preferred by 45% of respondents. This figure is in line with other studies conducted on the subject (Haddouche & Salomone, 2018; Kaihatu et al., 2021; Monaco, 2018), according to which the horizontal communication among tourists who exchange images, videos, news, advice and reviews through the web is assuming an increasingly central role in the younger generation and supplanting in many cases institutional communication channels.

To make the analysis more complete, the Mann-Whitney U-test was used to assess the gender difference in tourists' satisfaction levels (Mann & Whitney,

1947). The results indicate that the significance level for satisfaction by gender is greater than 5% (p=0.93), which means that with a reliability of 95%, there was no statistically significant difference in tourist satisfaction by gender of the respondents. Similarly, with a reliability of 95%, it can be also concluded that there is no statistically significant difference in tourist satisfaction considering the respondents' origin (p=0.92) and respondents' educational qualification (p=0.15). In this sense, it is safe to argue that the group of Gen Zers involved in the study appears quite compact and internally homogeneous.

Conclusions

Research findings allow the assertion that Gen Zers visited Vesuvius National Park between the years 2020 and 2021 mainly because they were fascinated by its environmental heritage. In this sense, it can be stated that their desire for ethics and eco-compatibility can be considered as a way to express their identity since their practical behaviours reflect their values. In fact, most of the Gen Z respondents stated that they lead an eco-sustainable everyday life. Consequently, even when they decide to plan their holidays, they almost always choose sustainable tourism, even if this sometimes implies having to spend a little more money.

Not only respect for the environment and the landscape but also interest in local people, their traditions and their customs can be defined as the main focus of Gen Z tourism, which is carried out as opposed to mass tourism. Their answers demonstrate their search for an alternative in the travelling experience to escape from the frenzy of everyday life. From this critical point of view, it is possible to argue that travel functions as a tool through which the members of the new generations can visit territories to discover the most meticulous details, enrich their identity, safeguard the environment and respect nature (Monaco, 2021; Sharpley, 2021; Stylos et al., 2021). This kind of journey can be considered a direct consequence of the attention that the younger generations are dedicating to environmental respect and to the preservation of landscapes. In other words, the attention paid to climate issues, the ecological transition, the global warming is also influencing tourism choices and the way of doing, living and interpreting travel (Corbisiero et al., 2022; Jaciow & Wolny, 2021; Lemy et al., 2019; Luttrell & McGrath, 2021; Mohr & Mohr, 2017).

The transition period we are experiencing towards climate neutrality cannot and must not consider only seeing the economic dimension. Today, more than ever, it seems central to combine the economic dimension with the social and environmental ones (Galgóczi, 2020; Jamal, 2019; McCauley & Heffron, 2018; Müller & Więckowski, 2018). Against this background, the members of the new generations through their increasingly sustainable behaviours are becoming the spokesman of a green change with a view to a complete and "just" transition, capable of making a change towards a less polluted world giving equal opportunities to all and reducing the negative environmental impacts from commercial activities (Monaco, 2022). One prediction can be made that, for example, in the tourism sector, accommodation facilities and restaurants with photovoltaic systems, recycled materials, local products or zero-km organic menus will represent in the short-medium term the main protagonists in the choices of the youngest tourists.

In the post-pandemic era, the reduction of travel has increased the chances of visiting places outdoors or travelling by means of green transportation, such as bicycles (Corbisiero & La Rocca, 2020; Huang et al., 2021; Marek, 2021). The world over the years has experienced various forms of pandemics and epidemics but none as widespread as COVID-19. Thus, the social and travelling crisis following the outbreak of the pandemic necessarily requires a revaluation and reorganisation of tourist flows. If mobility has suffered from several restrictions, the development of proximity tourism could become the new driving force so that the tourism sector restarts innovatively. The development of local tourism, for example, seems to represent a new trend capable of promoting tourism even without going too far from home (Corbisiero & Monaco, 2021; Hussain & Fusté-Forné, 2021; Seraphin & Dosquet, 2020; Umukoro et al., 2020). Emerging in the field of research, proximity tourism has also affected most Italian tourists, who have chosen to visit the paths of the Vesuvius National Park. In this sense, it could be one of the alternatives to travelling, in the medium-long term, since proximity tourism appears to be a way to contain the pandemic and the risk of contagion, also lessening the tourist load in the countries which suffer from overtourism and lack of sustainability (Fontanari & Traskevich, 2022; Ioannides & Gyimóthy, 2020; Koh, 2020; Milano et al., 2019). If this is the trend, today more than ever it is necessary that territories work to enhance local resources, giving greater prestige and visibility to their cultural, social, and environmental resources.

Although almost all of the participants in the study said they appreciated the visit to the Vesuvius National Park, they stressed that some of their expectations were not met. In particular, they would have liked to see more recycling bins, as well as the availability of more efficient public transportation. Instead, they had difficulties moving around in the Park, so they resorted to their own means of transport. This is the reason why the Park's public transport network was considered insufficient. In addition, many of the Park's paths have been defined as having little or no access, because they are located in places not easily reachable or poorly connected. In this sense, to improve the tourism experience in a sustainable way, it could be advisable to implement an electric transport network, which would facilitate the travelling of tourists and local inhabitants and exert less impact on the environment.

Furthermore, tourist enjoyment could also be improved through better communication with travellers. Some participants stated that they had difficulty finding information on the official Park's communication channels and had to resort to other sources. In other words, it would be necessary to create the possibility of direct, digitised, and fast communication channels.

These improvement goals could be achieved through the participation and involvement of the younger generations, for example, through the institution of consultation tables and participatory governance actions. Activating constant communication between young tourists and political decision-makers is of great necessity, as the former would express needs which can serve as the starting point for the latter to transform the tourist offer in a more sustainable and higher-quality way.

References

Andrade, G. S., & Rhodes, J. R. (2012). Protected areas and local communities: An inevitable partnership toward successful conservation strategies? *Ecology and Society, 17*(4), 14.

Bryman, A., & Bell, E. (2007). Business research methods. *Oxford International Journal of Social Research Methodology, 10,* 5–20.

Buongiorno, A., & Intini, M. (2021). Sustainable tourism and mobility development in natural protected areas: Evidence from Apulia. *Land Use Policy, 101,* 105220.

Bushell, R., & Bricker, K. (2017). Tourism in protected areas: Developing meaningful standards. *Tourism and Hospitality Research, 17*(1), 106–120.

Çaliskan, C. (2021). Sustainable tourism: Gen Z? *Journal of Multidisciplinary Academic Tourism, 6*(2), 107–115.

Corbisiero, F., & La Rocca, R. A. (2020). Tourism on demand: New form of urban and social demand of use after the pandemic event. *TeMA-Journal of Land Use, Mobility and Environment, 1,* 91–104.

Corbisiero, F., & Monaco, S. (2021). Post-pandemic tourism resilience: Changes in Italians' travel behavior and the possible responses of tourist cities. *Worldwide Hospitality and Tourism Themes, 13*(3), 401–417.

Corbisiero, F., Monaco, S., & Ruspini, E. (Eds.). (2022). *Millennials, Generation Z and the future of tourism.* Channel View Publications.

Cronbach, L. J. (1951). Coefficient alpha and the internal structure of tests. *Psychometrika, 16*(3), 297–334.

Dabija, D. C., Bejan, B., & Puşcaş, C. (2020). A qualitative approach to the sustainable orientation of Generation Z in retail: The case of Romania. *Journal of Risk and Financial Management, 13*(7), 152.

Dedeke, A. N. (2017). Creating sustainable tourism ventures in protected areas: An actor-network theory analysis. *Tourism Management, 61,* 161–172.

Dudley, N., Stolton, S., Belokurov, A., Krueger, L., Lopoukhine, N., MacKinnon, K., Sandwith, T., & Sekhran, N. (2010). *Natural solutions: Protected areas helping people cope with climate change.* IUCN/WCPA.

Dyer, M. I., & Holland, M. M. (1988). UNESCO's man and the biosphere program. *BioScience, 38*(9), 635–641.

Elsen, P. R., Monahan, W. B., Dougherty, E. R., & Merenlender, A. M. (2020). Keeping pace with climate change in global terrestrial protected areas. *Science Advances, 6*(25). https://doi.org/10.1126/sciadv.aay081

Entina, T., Karabulatova, I., Kormishova, A., Ekaterinovskaya, M., & Troyanskaya, M. (2021). Tourism industry management in the global transformation: Meeting the needs of Generation Z. *Polish Journal of Management Studies, 23*(2), 130–148.

Etikan, I., & Bala, K. (2017). Sampling and sampling methods. *Biometrics & Biostatistics International Journal, 5*(6), 00149.

Europarc Federation. (2010). *European charter for sustainable tourism in protected areas.* www.europarc.org/library/europarc-events-and-programmes/european-charter-for-sustainable-tourism/

Fermani, A., Sergi, M. R., Carrieri, A., Crespi, I., Picconi, L., & Saggino, A. (2020). Sustainable tourism and facilities preferences: The sustainable tourist stay scale (STSS) validation. *Sustainability, 12*(22), 9767.

Fontanari, M., & Traskevich, A. (2022). Smart-solutions for handling overtourism and developing destination resilience for the post-Covid-19 era. *Tourism Planning & Development.* https://doi.org/10.1080/21568316.2022.2056234

Galgóczi, B. (2020). Just transition on the ground: Challenges and opportunities for social dialogue. *European Journal of Industrial Relations*, *26*(4), 367–382.

Haddouche, H., & Salomone, C. (2018). Generation Z and the tourist experience: Tourist stories and use of social networks. *Journal of Tourism Futures*, *4*(1), 69–79.

Huang, S. S., Shao, Y., Zeng, Y., Liu, X., & Li, Z. (2021). Impacts of COVID-19 on Chinese nationals' tourism preferences. *Tourism Management Perspectives*, *40*, 100895.

Hussain, A., & Fusté-Forné, F. (2021). Post-pandemic recovery: A case of domestic tourism in Akaroa (South Island, New Zealand). *World*, *2*(1), 127–138.

International Union for Conservation of Nature (IUCN) World Commission on Protected Areas (WPCA) IUCN WCPA. (2018). *Guidelines for recognising and reporting other effective area-based conservation measures.* www.cbd.int/doc/c/0165/9fc3/962fae6c8e6 d0f8bc8ca361d/mcb-em-2018-01-inf-05-en.pdf

Ioannides, D., & Gyimóthy, S. (2020). The COVID-19 crisis as an opportunity for escaping the unsustainable global tourism path. *Tourism Geographies*, *22*(3), 624–632.

Jaciow, M., & Wolny, R. (2021). New technologies in the ecological behavior of Generation Z. *Procedia Computer Science*, *192*, 4780–4789.

Jamal, T. (2019). Tourism ethics: A perspective article. *Tourism Review*, *75*(1), 221–224.

Jiang, Y., & Hong, F. (2021). Examining the relationship between customer-perceived value of night-time tourism and destination attachment among Generation Z tourists in China. *Tourism Recreation Research*. https://doi.org/10.1080/02508281.2021.1915621

Kaihatu, T. S., Spence, M. T., Kasim, A., Gde Satrya, I. D., & Budidharmanto, L. P. (2021). Millennials' predisposition toward ecotourism: The influence of universalism value, horizontal collectivism and user generated content. *Journal of Ecotourism*, *20*(2), 145–164.

Koh, E. (2020). The end of over-tourism? Opportunities in a post-Covid-19 world. *International Journal of Tourism Cities*, *6*(4), 1015–1023.

Koulopoulos, T., & Keldsen, D. (Eds.). (2016). *Gen Z effect: The six forces shaping the future of business*. Routledge.

Laven, D. N., Wall-Reinius, S., & Fredman, P. (2015). New challenges for managing sustainable tourism in protected areas: An exploratory study of the European landscape convention in Sweden. *Society & Natural Resources*, *28*(10), 1126–1143.

Lazányi, K., & Bilan, Y. (2017). Generation Z on the labour market: Do they trust others within their work place? *Polish Journal of Management Studies*, *16*(1), 78–93.

Lemy, D. M., Hardianto, S. P., & Julita, Y. (2019). Generation Z and green hotel practices. *International Journal of Multidisciplinary Educational Research*, *8*(7), 309–331.

Locke, H., & Dearden, P. (2005). Rethinking protected area categories and the new paradigm. *Environmental Conservation*, *32*(1), 1–10.

Luttrell, R., & McGrath, K. (Eds.). (2021). *Gen Z: The superhero generation*. Rowman & Littlefield Publishers.

Maiorano, L., Falcucci, A., Garton, E. O., & Boitani, L. (2007). Contribution of the Natura 2000 network to biodiversity conservation in Italy. *Conservation Biology*, *21*(6), 1433–1444.

Mann, H. B., & Whitney, D. R. (1947). On a test of whether one of two random variables is stochastically larger than the other. *Annals of Mathematical Statistics*, *18*(1), 50–60.

Marek, W. (2021). Will the consequences of COVID-19 trigger a redefining of the role of transport in the development of sustainable tourism? *Sustainability*, *13*(4), 1887.

McCauley, D., & Heffron, R. (2018). Just transition: Integrating climate, energy and environmental justice. *Energy Policy*, *119*, 1–7.

Melillo, J. M., Lu, X., Kicklighter, D. W., Reilly, J. M., Cai, Y., & Sokolov, A. P. (2016). Protected areas' role in climate-change mitigation. *AMBIO: A Journal of the Human Environment, 45*(2), 133–145.

Milano, C., Cheer, J. M., & Novelli, M. (Eds.). (2019). *Overtourism: Excesses, discontents and measures in travel and tourism.* CABI Publishing.

Mohr, K. A., & Mohr, E. S. (2017). Understanding Generation Z students to promote a contemporary learning environment. *Journal on Empowering Teaching Excellence, 1*(1), 84–94.

Monaco, S. (2018). Tourism and the new generations: Emerging trends and social implications in Italy. *Journal of Tourism Futures, 4*(1), 7–15.

Monaco, S. (2021). *Tourism, safety and COVID-19: Security, digitization and tourist behaviour.* Routledge.

Monaco, S. (2022). Verso una transizione giusta? Sfide e prospettive socio-economiche della neutralità climatica. *Futuri, 17*, 99–110.

Müller, D. K., & Więckowski, M. (Eds.). (2018). *Tourism in transitions.* Springer.

Nunnally, J. C. (1978). An overview of psychological measurement. In B. B. Wolman (Ed.), *Clinical diagnosis of mental disorders* (pp. 97–146). Springer.

Paoletti, V., Secomandi, M., Piromallo, M., Giordano, F., Fedi, M., & Rapolla, A. (2005). Magnetic survey at the submerged archaeological site of Baia, Naples, Southern Italy. *Archaeological Prospection, 12*(1), 51–59.

Potts, T., Burdon, D., Jackson, E., Atkins, J., Saunders, J., Hastings, E., & Langmead, O. (2014). Do marine protected areas deliver flows of ecosystem services to support human welfare? *Marine Policy, 44*, 139–148.

Seraphin, H., & Dosquet, F. (2020). Mountain tourism and second home tourism as post COVID-19 lockdown placebo? *Worldwide Hospitality and Tourism Themes, 12*(4), 485–500.

Sharpley, R. (2021). On the need for sustainable tourism consumption. *Tourist Studies, 21*(1), 96–107.

Stone, M. T., & Nyaupane, G. P. (2016). Protected areas, tourism and community livelihoods linkages: A comprehensive analysis approach. *Journal of Sustainable Tourism, 24*(5), 673–693.

Stylos, N., Rahimi, R., Okumus, B., & Williams, S. (Eds.). (2021). *Generation Z marketing and management in tourism and hospitality.* Palgrave Macmillan.

Turner, A. (2015). Generation Z: Technology and social interest. *The Journal of Individual Psychology, 71*(2), 103–113.

Umukoro, G. M., Odey, V. E., & Yta, E. M. (2020). The effect of pandemic on home-based tourism: Post Covid-19. *International Journal of Humanities and Innovation, 3*(3), 115–120.

Wells, M. P., & McShane, T. O. (2004). Integrating protected area management with local needs and aspirations. *AMBIO: A Journal of the Human Environment, 33*(8), 513–519.

Williams, K., Page, R., & Petrosky, A. (2010). Multi-generational marketing: Descriptions, characteristics, lifestyles, and attitudes. *Journal of Applied Business and Economics, 11*(2). http://digitalcommons.www.na-businesspress.com/JABE/Jabe112/WilliamsWeb.pdf

Worboys, G. L., Lockwood, M., Kothari, A., Feary, S., & Pulsford, I. (Eds.). (2015). *Protected area governance and management.* Anu Press.

Yamane, T., & Kaneko, S. (2021). Is the younger generation a driving force toward achieving the sustainable development goals? Survey experiments. *Journal of Cleaner Production*, *292*, 125932.

Yussof, F. M., Harun, A., Norizan, N. S., Durani, N., Jamil, I., & Salleh, S. M. (2018). Retracted: The influence of social media consumption on Gen Z consumers' attitude. *Journal of Fundamental and Applied Sciences*, *10*(6S), 1288–1299.

Part III

Generation Z consumption behaviour in the hospitality sector

8 Generation Z tourists' perception of hotels' green practices

Lionel Saul and Cindy Yoonjoung Heo

Introduction

There is no doubt that green practices have become a critical issue within the global hotel industry (Han & Yoon, 2015), especially as hotels have a plethora of negative impacts on the environment. Indeed, hotels consume the largest amount of natural resources compared with all other types of commercial buildings (Filimonau et al., 2011) and produce significant quantities of solid waste (Radwan et al., 2012). Increased awareness of these issues is a positive step forward, especially because the tourism industry is such a visible industry, which makes it a possible catalyst of social change (Ryan, 2002).

With the prevailing global environmental crisis, hotel guests are becoming more environmentally conscious (Ham & Han, 2013). Therefore, more and more hoteliers acknowledge that hotels' active participation in green practices is an important condition for attracting and maintaining an increasing number of environmentally friendly customers. However, in spite of the significant impacts of hotels on the environment and hotel guests' growing awareness of hotels' green practices, some hoteliers are still cautious about going green (Kang et al., 2012; Pereira-Moliner et al., 2012), because hotels that seek to operate in a sustainable fashion must put up the initial outlay associated with going green (Bird et al., 2007; Zografakis et al., 2011) and need to analyse each product and process. While some studies have investigated which green practices are preferred by guests (Bryant, 2019; Manaktola & Jauhari, 2007; Millar et al., 2012; Trang et al., 2019), it is still challenging for hotels to implement the guests' preferred practices to ensure benefits for both the environment and the hotel. The challenge also lies in the fact that customers may consider if the quality of the rooms or the rates have been impacted by the implementation of green practices, which can result in a modification of the perceived service quality (Chia-Jung & Pei-Chun, 2014).

Generation Z (Gen Z) became an important consideration for the hotel industry as they move towards becoming a main market segment. Gen Z generally refers to individuals born since 1995 and comprises 32% of the world's population, more than Millennials, and its spending power continues to increase (Whitmore, 2019). According to Sheivachman (2017), this generation is willing to spend more time and money seeking pleasure and believes that travel is the most important thing in

DOI: 10.4324/9781003289586-11

their lives. Gen Z is driven by a cultural ethos of social justice and tries to get a sense of connection from the community (Meehan, 2016). Previous research has found that Gen Z is generally supportive of corporate social responsibility activities and green initiatives (e.g., Arıker & Toksoy, 2017). However, as the scope of hotels' green initiatives is wide and they include a vast array of practices, hoteliers need to prioritise the most important green practices that their guests care about.

Therefore, this chapter aims to fill a gap in the literature by exploring the green dimensions implemented by hotels that are the most important to Gen Z and how Gen Z evaluates a variety of green practices by adopting two measurement techniques (i.e., Likert scales and best-worst scaling [BWS]). Although the Likert scale is easily understood, and the responses are easily quantifiable, researchers have pointed out several inherent limitations associated with Likert scales such as the treatment of neutral opinions and acquiescence bias (e.g., Louviere et al., 1995; Chrzan & Skrapits, 1996; Cohen & Markowitz, 2002; Cohen & Neira, 2003). On the other hand, the BWS approach allows the most important attributes to be revealed by making respondents pick the most important and least important items from a choice set by making a trade-off between the alternatives (Cohen, 2009). In this chapter, both the Likert scale and BWS are used to identify the important green practices and dimensions that are considered to be of the greatest importance to Gen Z and the different results obtained by the two approaches are compared. The findings of this chapter should help hoteliers to develop adequate green practices to attract Gen Z, while reducing the environmental impact of their hotels.

Literature review

Hotels' green practices and guests' attitudes

A range of research (e.g., Dolnicar et al., 2019; Kallbekken & Sælen, 2013; Taylor et al., 2010) has discussed different approaches to green practices. For example, Kallbekken and Sælen (2013) discovered that "nudging" hotel guests can lead to reducing food waste by 20% without compromising guest satisfaction. Singh et al. (2014) showed that a hotel could reduce greenhouse gas emissions equivalent to 90 passenger vehicles annually by adequately recycling their waste. Filimonau and Magklaropoulou (2020) proposed to give financial incentives for guests to conserve energy during their stay while financial penalties could be applied for excessive energy use. Those studies underline the role that hotels can play in reducing the travel industry's greenhouse emissions.

Scholars have become interested in identifying factors that influence guests' willingness to stay in hotels with green practices. Consumers' environmental attitudes seem to impact favourably the decision to book a green hotel (Han et al., 2009; Kang et al., 2012). For example, the New Ecological Paradigm (NEP) has been widely adopted to measure environmental attitudes in previous literature (e.g., Dunlap et al., 2000), also in the hospitality industry field (Kang et al., 2012; Millar et al., 2012). Yet, some researchers suggest inconsistency between guests' environmental attitudes and their actual behaviours (e.g., Barber et al., 2012). Some studies explored which green practices are preferred by guests (Manaktola & Jauhari, 2007; Millar et al., 2012; Trang et al., 2019). For example, Trang et al. (2019) analysed 24 practices and regrouped them into five dimensions. However, most studies have compared only a few different green practices, which results in different rankings among the different studies.

Gen Z's attitude towards hotels' green practices

Gen Z has been discussed as a new sociological category, nurtured by the Internet, mobile technologies, and social media (Haddouche & Salomone, 2018). This generation already represents a significant proportion of all consumers worldwide and presents a tremendous opportunity for the tourism and hospitality industry to gain a better vision of how tourism demand could evolve in the near future. A hyper-connected generation, Gen Z have experienced considerable economic and social turmoil and grew up with challenges in relation to environmental sustainability and climate change, which have played a critical role in shaping their attitudes and beliefs, and travel preferences (Haddouche & Salomone, 2018). Previous research has found that Gen Z is generally supportive of corporate social responsibility activities and green initiatives (e.g., Arıker & Toksoy, 2017). According to the online Booking.com's (2019) report about Gen Z's travel consumptions, "56% would want to stay in a green or eco-friendly accommodation" and "54% state that the environmental impact travelling has on destinations is an important factor when deciding where to travel". As those individuals are growing in influence and spending power and comprise 32% of the world's population, being the largest group of consumers worldwide (Whitmore, 2019), they represent an opportunity for the industry. Especially, as 65% of Gen Z ranked "travel and seeing the world" (Booking.com, cited in Telus International, 2021) as the essential way to spend their money (Telus International, 2021). Thus, it is crucial for the industry to meet those individuals' expectations to attract this valuable population.

Methodology

Likert scale versus BWS approach

The Likert scale, an ordinal psychometric measurement of individuals' attitudes and opinions, is one of the most common scales used in social science studies, where responses to questions are measured on a continuum of two endpoints (Dittrich et al., 2007). The Likert scale is easily understood, and the responses are easily quantifiable. However, researchers have pointed out a number of inherent limitations associated with Likert scales. (e.g., Chrzan & Skrapits, 1996; Cohen & Markowitz, 2002; Cohen & Neira, 2003; Louviere et al., 1995). Although the Likert scale was proposed as an interval scale by assuming that two consecutive points are reflected within an equal distance in variation, respondents may not perceive the distances between two points of the scale the same way as others (Crask & Fox,

1987). To measure distinct levels of importance, several researchers have suggested adopting different types of scales (Cohen, 2009; Li, 2013).

One of the unique scales for rating the importance level of several attributes is BWS, proposed by Louviere and Woodworth (1983). The BWS approach allows the most important attributes to be revealed by making respondents pick the most important and least important items from a choice set by making a trade-off between the alternatives (Cohen, 2009). The statistical model at the basis of BWS is that the probability of a specified pair among a choice and an alternative is relative to the gap between the two attribute levels on the latent utility scale (Flynn et al., 2007). This approach has other advantages, such as preventing respondents from using the middle or one end of the scale as discrimination among items is forced (Cohen & Markowitz, 2002). Moreover, it allows a reduction in the impacts of individual idiosyncrasies that may arise from using rating scales due to cultural discrepancies or verbal ambiguities with labels (Lee et al., 2008). BWS is easy for participants to apprehend as there is only one way to select the most important and least items (Cohen & Markowitz, 2002). In this chapter, both the Likert scale and BWS are used to identify the important green practices which are appreciated by Gen Z and the different results obtained by the two approaches are compared.

Measurement

The 44 green practices studied in this chapter have been selected based on a review of previous literature (e.g., Filimonau & Magklaropoulou, 2020; Kallbekken & Sælen, 2013; Singh et al., 2014; Taylor et al., 2010; Trang et al., 2019), and the current practices of the industry and experts' suggestions. Based on Zhang et al.'s (2012) study, the 44 green practices were divided into two categories: a first one with guests being involved (i.e., requiring active participation from both the guests and the hotel operators), a second one with guests not being involved (i.e., only from the hotel operators). Then, inside both categories, the attributes were classified in dimensions regrouping the practices based on Trang et al. (2019). The objective of regrouping by dimension was to explore which area hoteliers should first focus on as well. By focusing on dimensions rather than only on practices, hoteliers could have a wider positive impact in terms of the preservation of the environment. Thus, five dimensions grouped the practices when guests are involved: "Application of green products and materials" (Product_1); "Waste reduction management" (Waste_1); "Energy management" (Energy_1); and "Water management and conservation" (Water_1); "Customers awareness" (Customers_1); and six when they are not: "Application of green products and materials" (Product_2); "Waste reduction management" (Waste_2); "Energy management" (Energy_2); and "Water management and conservation" (Water_2)"; "Protection of biodiversity" (Biodiversity_2); and "Employees awareness" (Employee_2).

In the first section of the survey, respondents were asked to assess the importance of each practice based on a 7-point Likert scale constructed as follows: "Not at all important"; "Low importance"; "Slightly important"; "Neutral"; "Moderately important"; "Very important"; "Extremely important". The second section was developed to assess among the dimension of green practices, which one is more important to the respondents, by using a BWS approach. One set contained the dimensions requiring participation from the guests, while the second those that do not require guests' participation. Respondents were asked to choose in both sets the most important and least important dimensions. In order to calculate the BWS score, the occurrence of the most and least important choices for each green dimension was cumulated. The BWS score was then calculated by subtracting the lowest score from its highest-scoring counterparts for each dimension. This score was divided by the number of respondents to represent the average best-worst (ABW). The best-worst (BW) ranking was determined by listing the dimensions according to the ABW scores.

Questions about respondents' profiles and the NEP scale were used to assess the respondents' attitudes towards the environment. In this chapter, the NEP is used as an internally consistent summated rating scale as used by Dunlap et al. (2000).

Data collection

Data were collected through an online survey. Of the 299 people who filled out the survey, 39 respondents needed to be removed as they did not complete the full survey. Among the valid answers, only 217 responses from members of Gen Z were analysed. In this chapter, Gen Z represents individuals born between 1995 and 2009.

Results

Profile of the respondents

The average age of participants was 22 years old: 50.7% (110 respondents) were male, 47.5% (103 respondents) were female and 1.8% (4) considered themselves as not belonging to the male or female category. The overall result of the NEP revealed a strong ecological endorsement (M=5.07, SD=.74) from Gen Z participants.

Results of the Likert scale and BWS approaches

As seen in Table 8.1, all the tested dimensions produced by the Likert scale have a score greater than 4.0 (Neutral), meaning that all the dimensions, whether they include the participation of the guests or not, are at least somewhat important to Gen Z. The green dimension Waste_1 (M=5.64, SD=1.35) had the highest score, followed by the dimension Customers_1 (M=4.86, SD=1.47), Energy_1 (M=4.77, SD=1.20), Water_1 (M=4.66, SD=1.67) and Product_1 (M=4.10, SD=1.16) for the ones requiring active participation from the guests and the hotel operators. Employee_2 (M=5.75, SD=1.57) had the highest score for the ones requiring active participation from the hotel operator only, followed by Waste_2 (M=5.28, SD=1.41), Water_2 (M=5.19, SD=1.39), Energy_2 (M=4.96, SD=1.24), Product_2 (M=4.88, SD=1.41), and Biodiversity_2 (M=4.66, SD=1.50).

Table 8.1 Generation Z's mean scores for hotel's green practices and dimensions (Likert scale)

Green practices requiring guest involvement	Mean	SD
Product_1[a]	**4.10**	**1.16**
Seasonal food on the menu only	5.07	1.72
Local food on the menu only	5.24	1.46
Vegetarian food on the menu only	2.82	1.83
Vegan food on the menu only	2.55	1.70
Food partially produced with onsite garden	4.53	1.58
Natural/cotton fibre for linen	4.51	1.67
Upcycled furniture	4.02	1.64
Waste_1[a]	**5.64**	**1.35**
Recycling bins in guests' rooms and common area	5.82	1.46
Reusable glass bottles and water fountains	5.68	1.49
Refillable shampoo, soap, and conditioner dispensers	5.41	1.58
Energy_1[a]	**4.77**	**1.20**
Appliances in sharing areas instead of in guests' room	3.77	1.63
Reuse policy for towel	5.40	1.55
Reuse policy for bedding	5.15	1.56
Incentives or penalties are applied whether the guest consumes less than the average guest's energy room consumption or more	4.78	1.74
Water_1[a]	**4.66**	**1.67**
Low-flow water tab	4.80	1.67
Low-flow water shower	4.53	1.83
Customers_1[a]	**4.86**	**1.47**
Hotel improves customers' knowledge of environmental responsibility	5.20	1.63
Savings achieved by the green practices displayed	4.94	1.70
Incentives encourage guests to travel in their own continents	4.45	1.84
Green practices not requiring guest involvement	Mean	SD
Product_2[a]	**4.88**	**1.41**
Bio-degradable material	5.70	1.62
Organic bathroom amenities	4.78	1.63
Hotel uses only upcycled furniture	4.29	1.63
Locally produced bathroom amenities	4.77	1.73
Waste_2[a]	**5.28**	**1.41**
Smaller plates at buffet to prevent overserving and a sign encouraging guests to help themselves more than once	4.84	1.75
Efficient recycling strategy	5.72	1.50
Energy_2[a]	**4.96**	**1.24**
Energy star rated appliances only	4.78	1.60
Wall insulation & triple glazing	5.04	1.56
Empty floors are closed during low occupancy period	5.07	1.66
Keycard master switches in guest rooms to control electricity usage	5.19	1.64
Ventilation heat recovery	4.97	1.55
LED for lightening	5.16	1.66
Sensors to monitor all the lights	5.17	1.59
Water is heated by solar thermal collectors	5.17	1.55
Electricity produces onsite by renewable energy	5.35	1.67
Renewable energy credit purchased	4.87	1.72
Daily tasks requiring high-power consumption are privileged when renewable energy is available in the electricity grid	4.88	1.61
Gas-powered tumble dryers	4.04	1.58

Green practices requiring guest involvement	Mean	SD
Efficient elevators	4.86	1.61
Water_2[a]	**5.19**	**1.39**
Rainwater harvesting for toilets is collected	5.09	1.56
Wastewater is collected for flushing toilets and irrigation	5.06	1.62
Water efficient toilets	5.43	1.53
Biodiversity_2[a]	**4.66**	**1.50**
Hotel possesses beehives	4.17	1.74
Conservation of local plants and animals onsite	5.14	1.61
Employee_2[a]	**5.75**	**1.57**
Employees are encouraged to be energy conscientious and to reduce waste	5.75	1.57

Note: n=217; scale ranged from 1=*Not at all important* to 7=*Extremely important*; M=*Mean*; SD=S*tandard deviation*
[a] Computed average of the practices' results that compose the dimension

Among all the dimensions, two stand out as they have practised with scores inferior to 4.0: Product_1 and Energy_1. Those practices are "Vegetarian food on the menu only" (M=2.82, SD=1.83), "Vegan food on the menu only" (M=2.55, SD=1.70.) and "Appliances in sharing areas instead of in guests' room" (M=3.77, SD=1.63). As a result, 41 out of the 44 tested green practices are supported by Gen Z.

The top three most important practices among the ones requiring active participation from the guests and hotel operators were "Recycling bins in guests' room and common area" (M=5.82, SD=1.46), "Reusable glass bottles and water fountains" (M=5.68, SD=1.49), "Refillable shampoo, soap and conditioner dispensers" (M=5.41, SD=1.58). The top three most important practices among the ones requiring active participation from the hotel operators only were "Employees are encouraged to be energy conscientious and to reduce waste" (M=5.75, SD=1.57), "Efficient recycling strategy" (M=5.72, SD=1.50), "Bio-degradable material" (M=5.70, SD=1.62).

The practices suggested by experts and tested for the first time in this type of research, to the best of the authors' knowledge, seemed to be important to be implemented for this generation: "Daily tasks requiring a high-power consumption are privileged when renewable energy is available in the electricity grid" (M=4.88, SD=1.61); "Incentives or penalties are applied whether the guest consumes less than the average guest's energy room consumption or more" (M=4.78, SD=1.74); "Incentives encourage guests to travel in their own continents" (M=4.45, SD=1.84).

The results of BWS revealed that when guests are involved, hoteliers should start by implementing first efficient Waste_1 and then Energy_1 programme, while when guests are not, Energy_2, Waste_2 and Water_2 programmes, as those five dimensions had an ABW score higher than zero (see Table 8.2).

When comparing the ranking of importance obtained with the Likert scale and the BWS, Waste_1 and Waste_2 obtained the first and second rank of importance, whether the guests are involved or not, regardless of the approach. The other dimensions had different ranks. Yet, other dimensions stand out due to their high

Table 8.2 Generation Z's best-worst scaling (BWS) scores for hotel's green dimensions

Green practices requiring guest involvement	Total best	Total worst	BW score	ABW score	Green practices not requiring guest involvement	Total best	Total worst	BW score	ABW score
Product_1	33	−58	−25	−.12	Product_2	28	−47	−19	−.09
Waste_1	67	−17	50	.23	Waste_2	43	−14	29	.13
Energy_1	49	−19	30	.14	Energy_2	52	−14	38	.18
Water_1	30	−33	−3	−.01	Water_2	24	−21	3	.01
Customers_1	38	−90	−52	−.24	Biodiversity_2	25	−74	−49	−.23
					Employee_2	45	−47	−2	−.01

Note: n=217; BW=best-worst, ABW=The average best-worst; positive signs on the BW and ABW scores indicate that the total number of dimensions chosen as the most important is greater than the total number of dimensions chosen as the least important. Negative signs indicate vice versa.

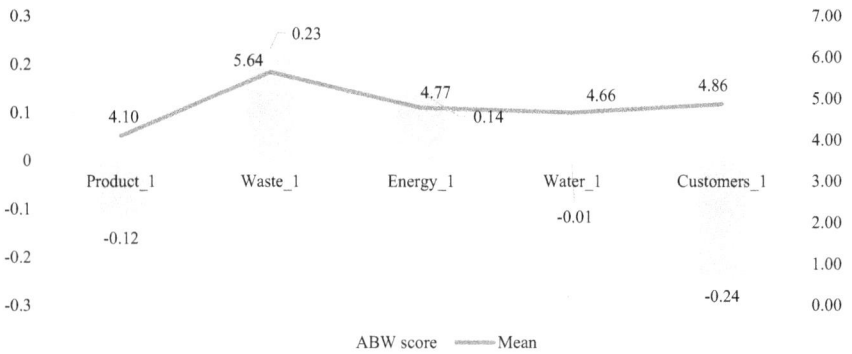

Figure 8.1 Ranking comparison of the Likert scale and best-worst scaling (BWS) requiring guest involvement

score with both approaches. Energy_1 ended up being the second most important with BWS, while third with the Likert scale. Energy_2 ended up being the first BWS, while fourth with the Likert scale. Employee_2 ended up being the fourth most important with BWS, while first with the Likert scale, see Figures 8.1 and 8.2.

Discussion and implications

Gen Z's responses to the 44 tested green practices are encouraging as only three of them were not rated as being at least somewhat important. This could be due to the apparent strong ecological endorsement of this generation. Participants were also supportive of the green practices that require their active participation. This is a particularly interesting finding as it seems that the loss in comfort and/or service quality that could be perceived from the implementation of the green practices requiring active participation from the guests does not override the willingness of Gen Z to encounter those practices in hotels. This finding is in opposition to the results of

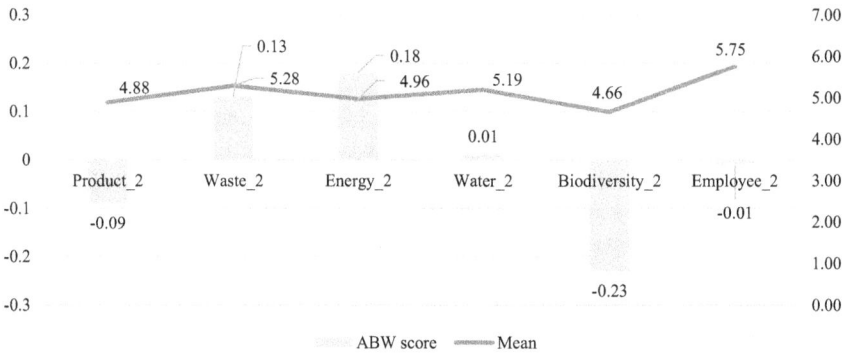

Figure 8.2 Ranking comparison of the Likert scale and best-worst scaling (BWS) not requiring guest involvement

Chia-Jung and Pei-Chun (2014), who stated that guests were less likely to accept those kinds of practices, such as having to bring their own toiletry. Thus, the results should reassure hoteliers that there are almost no wrong green practices to implement and that any green practices are likely to be well-received by Gen Z. By starting to implement the most preferred practices, hoteliers can now follow a step-by-step plan based on the results of this study to attract the eco-friendly Gen Z. Hoteliers can start an easy practice first such as locating recycling bins in guests' rooms and common areas and encourage the use of reusable glass bottles and water fountains. However, Gen Z doesn't seem ready to be too restricted when it comes to food, as vegetarian-only or vegan-only menus were not accepted. Those results could be explained by the fact that Gen Z, as with the rest of the population, still think that vegetarian or vegan diets are not enjoyable or too expensive (Bryant, 2019).

This chapter looked to go beyond the practice level and explored which dimensions were preferred by Gen Z, as implementing a few practices only would not be enough to mitigate the impact of the hotels on the environment. Tackling dimensions with as many of their related practices would be more significant but would require more investment from the hoteliers. Therefore, it is important for hoteliers to know which dimensions are the most preferred by Gen Z, so they can expect to attract those individuals while significantly reducing their ecological impact.

The comparison of the Likert scale and BWS revealed that hoteliers should start by implementing proper energy management and waste reduction programmes, and programmes to raise employees' awareness as those are the green dimensions most important for Gen Z, whether or not they are involved in such measures. However, as BWS forces trade-offs compared to the Likert scale (Lee et al., 2008), overall energy management and waste reduction programmes should be the top priorities to implement. Those results are partially consistent with the ones of Trang et al. (2019), who also found that an efficient energy management programme was one of the priorities of green consumers. It is worth mentioning that the most important dimensions, energy management and waste reduction programmes, have great potential in terms of greenhouse gas reduction according to the studies of Kall-bekken and Sælen (2013), Singh et al. (2014) and Taylor et al. (2010). Thus, the results of this chapter can be used by hoteliers to market and attract Gen Z, while significantly reducing the environmental impact of their hotels. Moreover, hoteliers should not forget that taking into account the environmental concern and environmental expectations of those individuals when travelling will prove to be beneficial in the long run as they are the largest group of consumers worldwide (Whitmore, 2019) and they are eager to travel (Telus International, 2021).

Gen Z seems to place a certain level of importance on the practices tested a priori for the first time with a Likert scale, which should stimulate hoteliers to think "outside the box" and review their business-as-usual model to be more innovative to reduce the ecological impact of hotels. Even though professionals seem to think that it is too early to implement practices that incentivise or penalise guests' energy consumption (Filimonau & Magklaropoulou, 2020), the findings of this chapter seem to show that Gen Z is willing to support such "pay-as-you-use" initiatives, which is consistent with the findings of Dolnicar et al. (2019). Moreover, the fact that Gen Z would seem to welcome incentives that encourage them to travel in their own continents underlines the role that hotels can play in reducing the travel industry's greenhouse emissions and participating in social change in the way people travel, which is consistent with the study of Ryan (2002).

Summary and conclusion

This chapter aims to explore the green dimensions implemented by hotels that are the most important to Gen Z and how Gen Z evaluates a variety of hotel green practices. The tested green practices covered many green dimensions: prioritising the use of green products and materials; waste reduction, energy management and water conservation programmes; protection of biodiversity; and raising customers' and employees' awareness. This chapter shows that Gen Z generally pays attention to hotels' green practices. Indeed, Gen Z's responses to the 44 tested green practices are encouraging as only three of them were not rated as being at least somewhat important. Moreover, the results of this study show that energy management and waste reduction programmes are the most important green dimensions for this generation, whose potential to reduce the negative impact of hotels on the environment has been proven. A surprising, yet promising finding is that Gen Z seems to believe that green "pay-as-you-use" practices are important.

Future studies may be conducted to generalise these results to other generations, locations, and segments as the respondents were mainly Gen Z in Switzerland. Future research may also further investigate how to positively communicate the efforts made by green hotels to help hoteliers market to Gen Z. Furthermore, this chapter revealed a strong ecological endorsement from Gen Z, but more research must be conducted to assess the link it may have in the selection of hotels with green practices.

References

Arıker, Ç., & Toksoy, A. (2017). Generation Z and CSR: Antecedents of purchasing intention of university students. *Journal of the Faculty of Economics and Administrative Sciences of Kafkas University, 8*(16), 483–502.

Barber, N., Kuo, P.-J., Bishop, M., & Goodman, R. (2012). Measuring psychographics to assess purchase intention and willingness to pay. *Journal of Consumer Marketing, 29*(4), 280–292.

Bird, R., Hall, A., Momente, F., & Reggiani, F. (2007). What corporate social responsibility activities are valued by the market? *Journal of Business Ethics, 76*(2), 189–206.

Booking.com. (2019). *GenZ unpacked Booking.com delves into the intentions of Gen Z travelers as they get ready to experience the world, their way*. https://destinationgenz.com

Bryant, C. J. (2019). We can't keep meating like this: Attitudes towards vegetarian and vegan diets in the United Kingdom. *Sustainability, 11*(23), 6844.

Chia-Jung, C., & Pei-Chun, C. (2014). Preferences and willingness to pay for green hotel attributes in tourist choice behavior: The case of Taiwan. *Journal of Travel & Tourism Marketing, 31*(8), 937–957.

Chrzan, K., & Skrapits, M. (1996). *Best-worst conjoint analysis: An empirical comparison with a full profile choice-based conjoint experiment*. INFORMS Marketing Science Conference.

Cohen, E. (2009). Applying best – worst scaling to wine marketing. *International Journal of Wine Business Research, 21*(1), 8–23.

Cohen, S. H., & Markowitz, P. (2002). *Renewing market segmentation: Some new tools to correct old problems* (pp. 595–612). ESOMAR 2002 Congress Proceedings.

Cohen, S. H., & Neira, L. (2003). *Measuring preference for product benefits across countries: Overcoming scale usage bias with maximum difference scaling*. ESOMAR 2003 Latin America Conference Proceedings.

Crask, M. R., & Fox, R. J. (1987). An exploration of the interval properties of 3 commonly used marketing-research scales: A magnitude estimation approach. *Journal of the Market Research Society, 29*(3), 317–339.

Dittrich, R., Francis, B., Hatzinger, R., & Katzenbeisser, W. (2007). A paired comparison approach for the analysis of sets of Likert-scale responses. *Statistical Modelling, 7*(1), 3–28.

Dolnicar, S., Cvelbar, L., & Grün, B. (2019). A sharing-based approach to enticing tourists to behave more environmentally friendly. *Journal of Travel Research, 58*(2), 241–252.

Dunlap, R. E., Van Liere, K. D., Mertig, A. G., & Jones, R. E. (2000). Measuring endorsement of the new ecological paradigm: A revised NEP scale. *Journal of Social Issues, 56*(3), 425–442.

Filimonau, V., Dickinson, J., Robbins, D., & Huijbregts, M. (2011). Reviewing the carbon footprint analysis of hotels: Life cycle energy analysis (LCEA) as a holistic method for carbon impact appraisal of tourist accommodation. *Journal of Cleaner Production, 19*(17–18), 1917–1930.

Filimonau, V., & Magklaropoulou, A. (2020). Exploring the viability of a new 'pay-as-you-use' energy management model in budget hotels. *International Journal of Hospitality Management, 89*, 102538.

Flynn, T. N., Louviere, J. J., Peters, T. J., & Coast, J. (2007). Best-worst scaling: What it can do for health care research and how to do it. *Journal of Health Economics, 26*(1), 171–189.

Haddouche, H., & Salomone, C. (2018). Generation Z and the tourist experience: Tourist stories and use of social networks. *Journal of Tourism Futures, 4*(1), 69–79.

Ham, S., & Han, H. (2013). Role of perceived fit with hotels' green practices in the formation of customer loyalty: Impact of environmental concerns. *Asia Pacific Journal of Tourism Research, 18*(7), 731–748.

Han, H., Hsu, L.-T., & Lee, J.-S. (2009). Empirical investigation of the roles of attitudes toward green behaviors, overall image, gender, and age in hotel customers' eco-friendly decision-making process. *International Journal of Hospitality Management, 28*(4), 519–528.

Han, H., & Yoon, H. J. (2015). Hotel customers' environmentally responsible behavioral intention: Impact of key constructs on decision in green consumerism. *International Journal of Hospitality Management, 45*, 22–33.

Kallbekken, S., & Sælen, H. (2013). "Nudging" hotel guests to reduce food waste as a win – win environmental measure. *Economics Letters, 119*(3), 325–327.

Kang, K. H., Stein, L., Heo, C. Y., & Lee, S. (2012). Consumers' willingness to pay for green initiatives of the hotel industry. *International Journal of Hospitality Management, 31*(2), 564–572.

Lee, J. A., Soutar, G., & Louviere, J. (2008). The best-worst scaling approach: An alternative to Schwartz's values survey. *Journal of Personality Assessment, 90*(4), 335–347.

Li, Q. (2013). A novel Likert scale based on fuzzy sets theory. *Expert Systems with Applications, 40*(5), 1609–1618.

Louviere, J. J., Swait, J., & Anderson, D. (1995). *Best/worst conjoint: A new preference elicitation method to simultaneously identify overall attribute importance and attribute level partworths* [Unpublished working paper]. University of Sydney.

Louviere, J. J., & Woodworth, G. (1983). Design and analysis of simulated consumer choice or allocation experiments: An approach based on aggregate data. *Journal of Marketing Research, 20*(4), 350–367.

Manaktola, K., & Jauhari, V. (2007). Exploring consumer attitude and behaviour towards green practices in the lodging industry in India. *International Journal of Contemporary Hospitality Management, 19*(5), 364–377.

Meehan, M. (2016). The next generation: What matters to gen we. *Forbes*. www.forbes.com/sites/marymeehan/2016/08/11/the-next-generation-what-matters-to-gen-we/#2565e7427350

Millar, M., Mayer, K. J., & Baloglu, S. (2012). Importance of green hotel attributes to business and leisure travelers. *Journal of Hospitality Marketing & Management, 21*(4), 395–413.

Pereira-Moliner, J., Claver-Cortés, E., Molina-Azorín, J., & Tarí, J. (2012). Quality management, environmental management and firm performance: Direct and mediating effects in the hotel industry. *Journal of Cleaner Production, 37*, 82–92.

Radwan, H. R., Jones, E., & Minoli, D. (2012). Solid waste management in small hotels: A comparison of green and non-green small hotels in Wales. *Journal of Sustainable Tourism, 20*(4), 533–550.

Ryan, C. (2002). Equity management, power sharing and sustainability-issues of the "new tourism". *Tourism Management, 23*(1), 17–26.

Sheivachman, A. (2017). U.S. millennials travel the most but Gen Z is on the rise. *Skift*. https://skift.com/2017/10/02/u-s-millennials-travel-the-most-but-gen-z-is-on-the-rise/

Singh, N., Cranage, D., & Lee, S. (2014). Green strategies for hotels: Estimation of recycling benefits. *International Journal of Hospitality Management, 43*, 13–22.

Taylor, S., Peacock, A., Banfill, P., & Shao, L. (2010). Reduction of greenhouse gas emissions from UK hotels in 2030. *Building and Environment, 45*(6), 1389–1400.

Telus International. (2021, June 3). *How Generation Z Is changing the future of travel.* www.telusinternational.com/articles/generation-z-future-of-travel

Trang, H. L., Lee, J.-S., & Han, H. (2019). How do green attributes elicit pro-environmental behaviors in guests? The case of green hotels in Vietnam. *Journal of Travel & Tourism Marketing, 36*(1), 14–28.

Whitmore, G. (2019). How Generation Z is changing travel for older generations. *Forbes.* www.forbes.com/sites/geoffwhitmore/2019/09/13/how-generation-z-is-changing-travel-for-older-generations/#760b74f278f7

Zhang, J. J., Joglekar, N., & Verma, R. (2012). Pushing the frontier of sustainable service operations management: Evidence from US hospitality industry. *Journal of Service Management, 23*(3), 377–399.

Zografakis, N., Gillas, K., Pollaki, A., Profylienou, M., Bounialetou, F., & Tsagarakis, K. (2011). Assessment of practices and technologies of energy saving and renewable energy sources in hotels in Crete. *Renewable Energy, 36*(5), 1323–1328.

9 Green practices and Gen Zers' behavioural intentions in the hospitality sector

Guadalupe Vila-Vazquez, Sandra Castro-González, and Belén Bande Vilela

Introduction

In recent years, a growing number of companies in the hospitality sector have become involved in initiatives that contribute to the protection and preservation of natural resources (Eid et al., 2021; Gao et al., 2016; Han, 2020; Kim & Hall, 2020; Verma et al., 2019; Vila-Vázquez et al., 2022; Yadav et al., 2016). In doing this, they try to respond to a growing trend of travellers looking for eco-friendly accommodation (Bordian et al., 2022; Han, 2021; Preziosi et al., 2019; Wang et al., 2018). In the academic field, this tendency has also given rise to the development of a new line of literature focused on green hotels, which are eco-friendly hotels that proactively seek to decrease their harmful environmental impact (Han, 2015).

Moreover, a new generation of travellers, Generation Z (Gen Z) – which comprises individuals born between the late 1990s and the late 2000s (Seemiller & Grace, 2019) – is becoming increasingly important in consumer decisions in the hospitality sector (European Travel Commission, 2020; Kamenidou et al., 2021; Robinson & Schänzel, 2019). Gen Zers not only are travellers of the future but also have an important prescriptive power (Monaco, 2018; Robinson & Schänzel, 2019). For this reason, understanding their concerns and interests is key to assessing future tourism demand, and responding appropriately to it.

Gen Zers are portrayed as highly sensitive to environmental issues, and as being positively disposed to sustainable tourism (European Travel Commission, 2020; Haddouche & Salomone, 2018). However, to the best of our knowledge, there is no evidence of this in the academic literature. In fact, contrary to expectations, the results of Haddouche and Salomone's (2018) study showed that sustainable tourism was not a key issue for the Gen Zers interviewed. To help clarify this issue, the main objective of this study is to analyse the effect of the implementation of environmentally responsible or irresponsible practices by hotels on Gen Zers' behavioural intentions. Two consumer behavioural intentions are considered in this study: the intention to stay in a hotel and the word-of-mouth (WOM) intention (Gao et al., 2016). We focus specifically on these variables because they correspond to the two main roles attributed to Gen Zers: their role as future consumers; and their role as influencers of green hotels.

DOI: 10.4324/9781003289586-12

Furthermore, based on the stimulus–organism–response (S-O-R) framework (Mehrabian & Russell, 1974), we also propose and test a mediating mechanism, specifically the hotel's corporate image. According to this model, the environmental stimuli (the environmentally responsible or irresponsible practices by the hotel) will influence behavioural intentions (the intention to stay in a hotel and the WOM intention) through its impact on the consumers' cognitive and affective reactions (in relation to the hotel's corporate image). Some research (e.g., Gao et al., 2016; Han, 2021; Martínez et al., 2019) has stated that consumer decision-making is largely affected by the set of beliefs or associations that consumers hold about a company. By studying the mediating role of the hotel's image, we contribute to addressing the shortage of studies that analyse the effects of consumers' perceptions of hotel companies. According to Gao et al. (2016), most studies about behavioural intentions in the hotel industry have focused on internalised perceptions "neglecting the importance of consumers' perceptions of the firms" (Gao et al., 2016, p. 113). Yadav et al. (2016) highlighted the point that despite the growing popularity of eco-friendly initiatives in the hospitality sector, few studies have analysed the effect of such practices on corporate image.

In addition, eco-friendly activities by the hotel are more likely to influence consumers' hotel image and their behavioural intentions if they are in line with their values (Chatzopoulou & de Kiewiet, 2021). Environmental or biospheric values – the values that emphasise the care of the environment and the biosphere – have been shown to have an important influence on the beliefs, personal norms, and behavioural intentions of green consumers (e.g., Han, 2020; Verma et al., 2019). However, to the best of our knowledge, we are unaware of any studies that analyse the moderating role between green practices and consumers' behavioural beliefs and intentions in the context of green hotels. Moreover, this issue is of special interest in the case of Gen Zers who have been characterised as highly interested in environmental issues (European Travel Commission, 2020).

Overall, following the S-O-R paradigm and the framework proposed by Gao et al. (2016), this study proposes that the effect of hotel involvement in environmentally responsible practices influences the behavioural intentions of Gen Zers through its impact on the perceptions of the company (the hotel image). In addition, the moderating effect of internalised perceptions (the environmental value) on the influence of green practices by the hotel, on both hotel image and consumers' behavioural intentions is studied. By examining this interactive effect, we extend knowledge about the conditions under which perceptions of hotels lead to a more favourable behavioural response among its targets.

This chapter is structured as follows. First, the development of the hypotheses is addressed, with a brief introduction to the theoretical framework underpinning the study model. Second, the methodology of the study is explained. Third, the results of the study are presented. Then, the results are discussed, the implications of the study are pointed out, and the main limitations and future research directions are indicated. Finally, the summary and conclusion section highlights the research questions that have been answered and the main findings of the research.

Hypothesis development

Stimulus–Organism–Response (S-O-R) framework

The S-O-R framework proposed by Mehrabian and Russel (1974) explains consumer behaviour through a three-step process: when individuals are exposed to an environmental stimulus (S); they generate internal evaluations (O); which prompt a response (R). The stimuli could be any external factor that influences customers' internal states (Bagozzi, 1986). Recent studies in the hotel sector have shown the stimulating role of variables related to eco-friendly practices by hotels in initiating the S-O-R process (Hameed et al., 2021; Su & Swanson, 2017; Wang et al., 2018). For example, Su and Swanson (2017), using the S-O-R framework, explained the effect of destination social responsibility on the pro-environmental behaviour of Chinese tourists through consumption emotions and tourist-destination identification. Similarly, Wang et al. (2018) examined the relationship between a hotel's green image and consumers' WOM intention via consumers' green trust and green satisfaction, under the perspective of the S-O-R paradigm. Recently, Hameed et al. (2021) used this framework to assess the effect of green practices on customers' green WOM intention for hotels with environmentally responsible operations through green image, green satisfaction, and green trust.

In our model, it is proposed that the environmentally responsible or irresponsible practices by the hotel (Stimuli) evoke a positive hotel image (Organism), which then promotes positive behavioural intentions (Response).

Green practices, hotel image, and consumer behavioural intentions

Hotel involvement in green practices – namely the reduction of consumption of water, waste, and energy, recycling, and so on – is a prerequisite for attracting and retaining environmentally conscious travellers (Han, 2020; Kim et al., 2017). However, as consumer behavioural intentions are a complex phenomenon that can be influenced by multiple variables, direct influence is not always apparent (Boccia et al., 2019; Ramesh et al., 2019). For example, in the study carried out by Martínez et al. (2019), it was shown that the hotel's green image did not directly influence behavioural intentions, but that its effect was produced through its impact on the hotel's corporate image.

According to Ramesh et al. (2019), corporate social responsibility (CSR) on consumers' perceptions have a major role to play in generating valuable content for brand image construction. Green activities by the hotel show its concern for the environment and its willingness to strive for the good of society, generating a positive image of the hotel in the mind of the consumer (Eid et al., 2021; Yadav et al., 2016). This positive image strongly influences the consumer decision-making process "because this asset offers mental shortcuts when processing information" (Martínez et al., 2019, p. 1384). Lee et al. (2010) showed that the better the perception of the hotel by the consumers, the more positive their behavioural intentions will be (i.e., intention to revisit, intent to spread positive WOM, and willingness to pay a premium). In addition, the meta-analysis carried out by Gao et al. (2016) shows a significant positive influence of company image on consumer behavioural intentions.

Behavioural intentions reflect the probability of consumer involvement in a particular behaviour (Ajzen & Fishbein, 1980). Thus, when behavioural intentions are favourable, there is a greater likelihood that the customer will stay at the hotel or speak well of it (Yadav et al., 2016). Indeed, Eid et al. (2021) reported a very strong relationship between the guests' intention to visit green hotels and their actual purchase ($\beta = 0.79$). Moreover, in the hotel industry, there is some evidence of the mediating role of corporate image in the relationship between green practices and consumer behavioural intentions (e.g., Hameed et al., 2021; Martínez et al., 2019; Yadav et al., 2016). Based on the aforementioned arguments and previous evidence, the following hypotheses are proposed:

H1: Hotel involvement in green practices will indirectly relate to consumers' purchase intention through the image of the hotel.

H2: Hotel involvement in green practices will indirectly relate to WOM intention through the image of the hotel.

The moderating role of environmental value

Environmental value has been recognised as a relevant variable in the value-belief-norm theory (Stern, 2000), as well as in the recent theory of green purchase behaviour (Han, 2020). According to these theories, green purchase intention/behaviour is determined by the sense of moral obligation experienced by consumers (personal norm), which is activated by the influence of environmental values in the ecological world view awareness of consequences-ascription of responsibility sequential process (Han, 2015, 2020). Along these lines, Eid et al. (2021) corroborated that the sense of being obliged to visit green hotels is strongly and positively influenced by the guests' environmental values. Likewise, the study by Verma et al. (2019), conducted in the hotel sector in India, evidenced the relevance of environmental values in predicting consumers' environmental concern and attitude towards green hotels, which in turn determined their intention to visit a green hotel.

The aforementioned relationships between values and various pro-environmental attitudes and behavioural intentions seem to suggest that guests' environmental values will condition the influence of the hotels' green initiatives in their purchase decision processes. That is, when consumers show a high interest in environmental issues, the hotel's environmental practices (both responsible and irresponsible) will have a greater value in the formation of the hotel's image, which will indirectly influence consumers' behavioural intentions towards the hotel. If consumers show less interest in these issues, the hotel's image will be more influenced by other traditional attributes, and the influence of hotel green practices on both hotel image and behavioural intentions will be less.

A study by Amatulli et al. (2021) showed that consumers' environmental concerns strengthened the relationship between the hotel's sustainability-focused communication and the perception of hotel integrity. They also showed that a hotel's sustainability-focused communication indirectly influenced willingness to book a room through perceived hotel integrity when guest environmental concern was medium or high. Based on the aforementioned reasoning and the previous evidence, the following hypotheses are proposed:

H3: The higher the level of guests' environmental values, the greater the impact of the green practices on the hotel's brand image.

H4: The higher the level of guests' environmental values, the greater the indirect impact of the green practices on the hotel's purchase intention via the hotel's brand image.

H5: The higher the level of guests' environmental values, the greater the indirect impact of the green practices on the hotel's WOM intention via the hotel's brand image.

Method

Following the example of Amatulli et al. (2021) a two-cell (eco-friendly vs. non-eco-friendly hotel) between-subjects experiment was carried out with 155 voluntary participants. Their mean age was 19.26 years (SD=1.596) and 48% were female. All were undergraduate students from the business world in Spain.

First, participants assessed their environmental value – also known as biosphere value – using the scale employed by Verma et al. (2019), which includes the items from Stern et al.'s (1999) altruistic values scale concerning the environment. The scale created by Stern et al. (1999) in the development of their 'value-belief-norm theory' is widely used to measure environmental value (e.g., Eid et al., 2021; Yadav et al., 2019). Next, they were asked to imagine that they were looking for a hotel to stay in for a city break, and they were provided with a description of a fictitious hotel. Approximately half of the participants (80) read a description of a hotel committed to environmentally responsible practices, highlighting information such as policies to reduce energy and water consumption, recycling, and responsible waste management. The other half (75 participants) read a description of a hotel with all the amenities, but that was criticised for its misuse of water and energy. They were then asked to rate their perception of the hotel's brand image, their purchase intention, and their WOM intention. The scales used by Lien et al. (2015) were adapted to assess both hotel brand image and purchase intention. Consumers' WOM intention was measured with the scale employed by Wang et al. (2018). Finally, they were asked to report their demographic data (age and sex).

Analysis of the data was carried out with SPSS Statistics 25 and SPSS AMOS 23. Firstly, a confirmatory factor analysis (CFA), using AMOS, was conducted on a total sample of 155 participants to corroborate the reliability and validity of the constructs used in the study. Secondly, to contrast the hypotheses a mediation analysis and a moderated mediation analysis were carried out using the PROCESS macro for SPSS (Hayes, 2017). This tool allows the analysis of the mediation and moderation relationships over the total sample, using multiple regression by ordinary least squares.

Results

Table 9.1 shows descriptive statistics and correlations. Additionally, Cronbach's alpha values are shown in the diagonal axis. Moreover, the last two columns show the composite reliability (CR) and the average variance extracted (AVE) values.

The implementation of environmentally responsible practices by the hotel was captured by a binary variable called the eco-friendly hotel that took the value 1 if the hotel was involved in environmentally responsible practices and 0 otherwise.

Results of the CFA revealed an acceptable fit of the measurement model: ($\chi2$ (83) = 149.657; $\chi2$/df =1.803; CFI =0.971; IFI = 0.971; RMSEA = 0.072). All the factor loadings are significant and higher than the ideal value of 0.7, supporting the constructs' convergent validity. As is shown in Table 9.1, the measures' reliability was also confirmed since the CR and the AVE were above the recommended thresholds (Hair et al., 2010). Furthermore, the constructs also showed discriminant validity, as the inter-constructs correlation confidence intervals (CI) excluded the unit value and their squared values were lower than AVE (Hair et al., 2010).

Then the study hypotheses were tested. Model 4 from Hayes (2017) was used to analyse the first two hypotheses, and model 7 was used to study the moderation hypothesis (hypothesis 3) and the moderate mediation hypothesis (hypotheses 4 and 5).

The results of model 4 are presented in Figure 9.1. As can be seen, the consideration of a hotel as eco-friendly was positively and significatively associated with the hotel's image (a_1 = 1.10; p < 0.001); in turn, the hotel's image positively and significantly influences both purchase intention (b_{1a} = 0.64; p < 0.001) and WOM intention (b_{1b} = 0.51; p < 0.001). Moreover, the confidence intervals for the

Table 9.1 Descriptive statistics, correlations, and Cronbach's alpha values

Variable	M	SD	1	2	3	4	5	6	7	AVE	CR
1. Gender	–	–									
2. Age	19.260	1.596	−0.129								
3. Eco-friendly hotel	–	–	−0.018	−0.001							
4. Hotel's image	5.115	1.287	−0.021	−0.037	0.407***	(0.836)				0.640	0.842
5. Purchase intention	4.274	1.331	−0.108	−0.148	0.361***	0.686***	(0.934)			0.783	0.935
6. WOM intention	4.313	1.642	−0.114	−0.087	0.572***	0.699***	0.029	(0.967)		0.875	0.965
7. Environmental value	5.807	1.069	0.146	0.052	−0.161*	0.044	0.029	−0.042	(0.876)	0.656	0.884

Note: N = 155. *p < 0.05, **p < 0.01, ***p < 0.001. CR, composite reliability; AVE, average variance extracted.

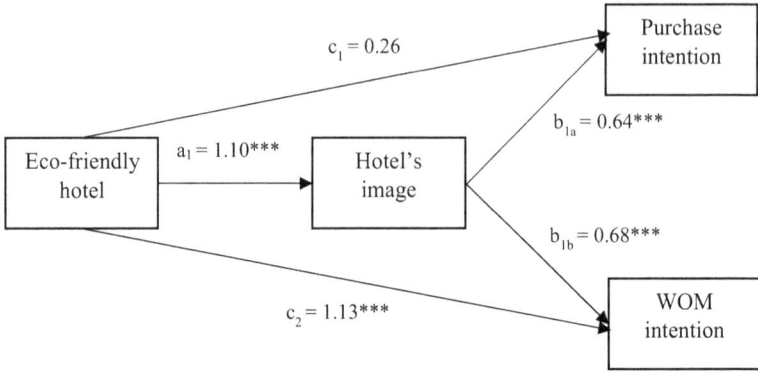

Figure 9.1 Unstandardised coefficients for the mediation model (Model 4)
Note: ***p < 0.001.

Table 9.2 Non-standardised coefficients for the moderated mediation model (Model 7)

Variables	*Model 1. Hotel's Image*				*Model 2a. Hotel's Purchase Intention*				*Model 2b. WOM Intention*			
	Coeff	p	LLCI	ULCI	Coeff	p	LLCI	ULCI	Coeff	p	LLCI	ULCI
Constant	5.09	<0.001	4.90	5.29	1.06	0.001	0.42	1.71	0.31	0.358	−0.35	0.97
Eco-friendly Hotel	1.12	<0.001	0.74	1.51	0.26	0.131	−0.08	0.60	1.13	<0.001	0.76	1.50
Environmental value	0.09	0.332	−0.09	0.27								
Interaction	0.53	0.005	0.16	0.90								
Hotel's Image					0.64	<0.001	0.51	0.76	0.68	<0.001	0.54	0.82
R^2		0.217				0.478				0.587		
F		13.950				69.667				108.056		
p value		<0.001				<0.001				<0.001		

Note: N = 155; Interaction = Eco-friendly hotel × Environmental value.

indirect effect estimated by bootstrap exclude the zero value (purchase intention 95 % CI = 0.419; 1.032; WOM intention 95 % CI = 0.441; 1.105). This supports hypotheses 1 and 2, respectively.

In Table 9.2, the results of model 7 are shown. In this model, the variables involved in the interaction term were mean-centred to facilitate the interpretability of the coefficients (Hayes, 2017). As shown, environmental values moderate the impact of the consideration of a hotel as eco-friendly in the terms of the hotel's image (a_{31} = 0.53; p= 0.005), supporting hypothesis 3. Finally, the confidence intervals for the index of moderated mediation do not include zero (purchase intention 95 % CI = 0.087; 0.630; WOM intention 95 % CI = 0.089; 0.638), providing support for hypotheses 4 and 5 (Hayes, 2015).

The study of moderator values defining Johnson–Neyman significance regions provides us with valuable information about the relevance of environmental values in the Gen Zers. Specifically, this analysis shows that the influence of the implementation of green practices by the hotel only influences the hotel's image from medium – high values of environmental value (specifically from 4.745). However, 84.52% of the sample presented values higher than this cut-off value. This seems to corroborate the high standards of prosocial values in Gen Z.

Discussion

Results, discussion, and implications

Despite the importance of Gen Zers for the travel industry, there is still a lack of studies that analyse Gen Z as travellers (European Travel Commission, 2020; Kamenidou et al., 2021; Robinson & Schänzel, 2019). Gen Zers "will be the most active players in the tourism market" (Monaco, 2018, p. 9). In addition, they also have an important influence on the purchasing decisions of both their family and third parties (Kamenidou et al., 2021; Robinson & Schänzel, 2019). Furthermore, their predisposition to use social networks causes them to spread e-WOM about the tourist services they use, thus reaching a wide target audience (Clark et al., 2021; Kamenidou et al., 2021). Therefore, understanding the interests and motivations of Gen Zers in terms of their hospitality choices is essential for the management of hotel companies, and for them to be able to implement strategies that guarantee their survival and growth. In this sense, the main objective of this research focused on analysing the impact of environmentally responsible/irresponsible practices on the behavioural intentions of Gen Zers in the hotel sector.

The results obtained show, in general, that the hotel's environmental performance affects its corporate image, which in turn determines Gen Zers' behavioural intentions about the hotel. The perception of a hotel as eco-friendly by Gen Zers leads them not only to avoid the punishment attached to disregarding the protection of the environment but also to have a more positive image about it, while if the hotel is perceived as non-eco-friendly its corporate image will be worse. Furthermore, this finding is in line with the results of previous studies in the hotel context among a general generational sample (Eid et al., 2021; Martínez et al., 2019; Ramesh et al., 2019; Yadav et al., 2016). In addition, the study results are consistent with the S-O-R framework (Mehrabian & Russel, 1974). The hotel's environmental performance acts as a stimulus that evokes a positive image of the hotel (organism), which in turn, promotes a positive response in the form of increased intention to hotel's purchase and speak well of it. The results of the study also show that, while the influence of hotel environmental performance on the intention to stay in a hotel was fully mediated by hotel image, the mediation was partial when WOM intention is the outcome. This seems to indicate the need to consider other mediating variables in this relationship such as, for example, the perception of the hotel's trust or integrity.

Moreover, this study showed that the impact of a hotel's environmental practices on its image depends on the values of its potential customers. Customers create an image of the hotel considering its values, strategies, and competencies. However, the guests' values influence the weight they give to the different attributes of the hotel when generating their image of it. While there is some evidence for the antecedent role of environmental value in the green purchase/intention through the values-belief-norm chain (e.g., Verma et al., 2019), to the best of our knowledge this is the first study to analyse its moderating role in the relationship between an environmental stimulus and a consumer' cognitive and affective reaction. In this sense, we deepen the knowledge of the S-O-R framework by including the moderating role of consumer values. Specifically, this study shows that in Gen Zers, among guests with high environmental values, the environmental practices carried out by the hotel have a significant impact on the hotel's image. Consequently, the results of this research support the premise that Gen Zers are characterised by their interest in protecting and caring for the environment (European Travel Commission, 2020; Haddouche & Salomone, 2018).

From a practical perspective, this study confirms eco-friendly practices as strategic initiatives in the hospitality sector, as reflecting the hotel's concern for the environment generates a more positive image of the hotel in the minds of Gen Zers. This, in turn, generates a greater intention to stay at the hotel, and more positive communication (WOM) from Gen Zers. Consequently, hotel management must articulate clear messages about their sustainability efforts, and convey them through the right channels to reach Gen Zers, as their ability to compete will depend on it (Clark et al., 2021).

According to Martínez et al. (2019), the use of multiple channels in an integrated way facilitates the communication of the hotels' environmental positioning. Possible channels include: "media advertising, speeches, reports, press releases, web pages, announcements, newsletters, articles, and corporate social responsibility reports", in addition to, "social media channels" (Martínez et al., 2019, p. 1388). However, recent studies (e.g., Eid et al., 2021) show that although green activities significantly influence the corporate image, the hotel's communication of these practices does not improve its image when it is carried out through the traditional marketing mix media (i.e., advertisement, public relations activities, sales promotion, and sponsorship activities). On the contrary, the results of the recent study by Clark et al. (2021) supported the usefulness of the use of social media marketing by hotels to communicate these initiatives in increasing intentions to stay and expanding eWOM by young travellers (including Millennials and Gen Zers). For their part, Kapoor et al. (2021) demonstrated that the effectiveness of positive and negative emotional appeals in sustainability communication in social media depends on the source of the message. The results of their study concluded that messages posted by third parties, such as social media influencers, are more effective when they have the guilt appeal, while the hotel's own social media messages are more persuasive when they have the sensual appeal.

Another effective way to communicate the hotel's involvement in environmental initiatives to guests is through environmental certifications (Martínez et al., 2019). In addition, these make social media marketing communications more reliable, allowing to differentiate hotels with genuine concern for environmental issues from 'green-washed' hotels (Daddi et al., 2019). The main alternatives for environmental certification include the Green Business Certification, the ISO 14001 or the ECO-Management and Audit Scheme.

Limitations of the study and future research

Despite its contributions, this study is not without limitations. First, this study has focused on a sample of Gen Zers in Spain which precludes the generalisation of results. Future studies in other cultural contexts are needed. Second, the consideration of other mediators within the S-O-R framework would allow the elucidation of new mechanisms of influence. Other relevant behavioural intentions, such as the predisposition to pay more for staying in a green hotel, could also be studied. Finally, the study of the effectiveness of different communication channels of hotels' sustainability commitments for Gen Zers is a field of research of great interest.

Summary and conclusion

Although the literature has recognised the important role of sustainability efforts by hotels in developing positive behavioural intentions in travellers, there is a need to study their effectiveness among Gen Zers. The relevance of such analysis is due not only to their role as future travellers but also to their role as prescribers. Results indicated that green practices influence Gen Zers' behavioural intentions through their impact in relation to a hotel's image. In addition, the consumers' environmental values moderated the green practice-hotel image-behavioural intentions relationship. That is, consumers' environmental values condition the weight that hotel green practices have on their hotel's image. Thus, the greater the interest in the care and protection of the environment by the consumer, the stronger the aforementioned relationship will be. This research contributes to the literature on green hotels by expanding knowledge about an important and largely unexplored market segment. Likewise, at the theoretical level, the findings of this study, in line with the S-O-R framework, respond to the call for new theories (beyond the theories of planned behaviour or reasoned action) that explain the influence of environmentally responsible initiatives on the behavioural intentions of guests (Gao et al., 2016). It also provides useful insight to hotel management to enhance and promote the intention to stay in hotels, as well as WOM intention among Gen Zers.

References

Ajzen, I., & Fishbein, M. (Eds.). (1980). *Understanding attitudes and predicting social behavior*. Prentice Hall.

Amatulli, C., De Angelis, M., & Stoppani, A. (2021). The appeal of sustainability in luxury hospitality: An investigation on the role of perceived integrity. *Tourism Management, 83*, 104228.

Bagozzi, R. P. (1986). *Principles of marketing management*. Science Research Associates.

Boccia, F., Malgeri Manzo, R., & Covino, D. (2019). Consumer behavior and corporate social responsibility: An evaluation by a choice experiment. *Corporate Social Responsibility and Environmental Management*, *26*(1), 97–105.

Bordian, M., Gil-Saura, I., & Šerić, M. (2022). ¿Cómo impulsa la Comunicación Integrada de Marketing la satisfacción del huésped? Una propuesta a través del conocimiento ecológico y la co-creación de valor. *Cuadernos De Gestión*, *22*(1), 7–20.

Chatzopoulou, E., & de Kiewiet, A. (2021). Millennials' evaluation of corporate social responsibility: The wants and needs of the largest and most ethical generation. *Journal of Consumer Behaviour*, *20*(3), 521–534.

Clark, M., Kang, B., & Calhoun, J. R. (2021). Green meets social media: Young travelers' perceptions of hotel environmental sustainability. *Journal of Hospitality and Tourism Insights*. https://doi.org/10.1108/JHTI-03-2021-0062

Daddi, T., Ceglia, D., Bianchi, G., & de Barcellos, M. D. (2019). Paradoxical tensions and corporate sustainability: A focus on circular economy business cases. *Corporate Social Responsibility and Environmental Management*, *26*(4), 770–780.

Eid, R., Agag, G., & Shehawy, Y. M. (2021). Understanding guests' intention to visit green hotels. *Journal of Hospitality and Tourism Research*, *45*(3), 494–528.

European Travel Commission. (2020). *Study on Generation Z travellers*. https://etc-corporate.org/uploads/2020/07/2020_ETC-Study-Generation-Z-Travellers.pdf

Gao, Y. L., Mattila, A. S., & Lee, S. (2016). A meta-analysis of behavioral intentions for environment-friendly initiatives in hospitality research. *International Journal of Hospitality Management*, *54*, 107–115.

Haddouche, H., & Salomone, C. (2018). Generation Z and the tourist experience: Tourist stories and use of social networks. *Journal of Tourism Futures*, *4*(1), 69–79.

Hair, J. F., Anderson, R. E., Babin, B. J., & Black, W. C. (Eds.). (2010). *Multivariate data analysis: A global perspective*. Pearson Prentice Hall.

Hameed, I., Hussain, H., & Khan, K. (2021). The role of green practices toward the green word-of-mouth using stimulus-organism-response model. *Journal of Hospitality and Tourism Insights*. https://doi.org/10.1108/JHTI-04-2021-0096

Han, H. (2015). Travelers' pro-environmental behavior in a green lodging context: Converging value-belief-norm theory and the theory of planned behavior. *Tourism Management*, *47*, 164–177.

Han, H. (2020). Theory of green purchase behavior (TGPB): A new theory for sustainable consumption of green hotel and green restaurant products. *Business Strategy and the Environment*, *29*(6), 2815–2828.

Han, H. (2021). Consumer behavior and environmental sustainability in tourism and hospitality: A review of theories, concepts, and latest research. *Journal of Sustainable Tourism*, *29*(7), 1021–1042.

Hayes, A. F. (2015). An index and test of linear moderated mediation. *Multivariate Behavioral Research*, *50*(1), 1–22.

Hayes, A. F. (2017). *Introduction to mediation, moderation, and conditional process analysis: A regression-based approach*. Guilford Publications.

Kamenidou, I., Vassilikopoulou, A., & Priporas, C.-V. (2021). New Sheriff in town? Discovering Generation Z as tourists. In N. Stylos, R. Rahimi, B. Okumus, & S. Williams (Eds.), *Generation Z marketing and management in tourism and hospitality* (pp. 121–140). Palgrave Macmillan.

Kapoor, P. S., Balaji, M. S., & Jiang, Y. (2021). Effectiveness of sustainability communication on social media: Role of message appeal and message source. *International Journal of Contemporary Hospitality Management, 33*(3), 949–972.

Kim, M. J., & Hall, C. M. (2020). Can sustainable restaurant practices enhance customer loyalty? The roles of value theory and environmental concerns. *Journal of Hospitality and Tourism Management, 43*, 127–138.

Kim, W. G., Li, J. J., Han, J. S., & Kim, Y. (2017). The influence of recent hotel amenities and green practices on guests' price premium and revisit intention. *Tourism Economics, 23*(3), 577–593.

Lee, J. S., Hsu, L. T., Han, H., & Kim, Y. (2010). Understanding how consumers view green hotels: How a hotel's green image can influence behavioural intentions. *Journal of Sustainable Tourism, 18*(7), 901–914.

Lien, C. H., Wen, M. J., Huang, L. C., & Wu, K. L. (2015). Online hotel booking: The effects of brand image, price, trust and value on purchase intentions. *Asia Pacific Management Review, 20*(4), 210–218.

Martínez, P., Herrero, Á., & Gómez-López, R. (2019). Corporate images and customer behavioral intentions in an environmentally certified context: Promoting environmental sustainability in the hospitality industry. *Corporate Social Responsibility and Environmental Management, 26*(6), 1382–1391.

Mehrabian, A., & Russell, J. A. (Eds.). (1974). *An approach to environmental psychology.* The MIT Press.

Monaco, S. (2018). Tourism and the new generations: Emerging trends and social implications in Italy. *Journal of Tourism Futures, 4*(1), 7–15.

Preziosi, M., Tourais, P., Acampora, A., Videira, N., & Merli, R. (2019). The role of environmental practices and communication on guest loyalty: Examining EU-ecolabel in Portuguese hotels. *Journal of Cleaner Production, 237*, 117659.

Ramesh, K., Saha, R., Goswami, S., Sekar, & Dahiya, R. (2019). Consumer's response to CSR activities: Mediating role of brand image and brand attitude. *Corporate Social Responsibility and Environmental Management, 26*(2), 377–387.

Robinson, V. M., & Schänzel, H. A. (2019). A tourism inflex: Generation Z travel experiences. *Journal of Tourism Futures, 5*(2), 127–141.

Seemiller, C., & Grace, M. (Eds.). (2019). *Generation Z: A century in the making.* Routledge.

Stern, P. C. (2000). New environmental theories: Toward a coherent theory of environmentally significant behavior. *Journal of Social Issues, 56*(3), 407–424.

Stern, P. C., Dietz, T., Abel, T., Guagnano, G. A., & Kalof, L. (1999). A value-belief-norm theory of support for social movements: The case of environmentalism. *Human Ecology Review, 6*(2), 81–97.

Su, L., & Swanson, S. R. (2017). The effect of destination social responsibility on tourist environmentally responsible behavior: Compared analysis of first-time and repeat tourists. *Tourism Management, 60*, 308–321.

Verma, V. K., Chandra, B., & Kumar, S. (2019). Values and ascribed responsibility to predict consumers' attitude and concern towards green hotel visit intention. *Journal of Business Research, 96*, 206–216.

Vila-Vázquez, G., Castro-Casal, C., & Carballo-Penela, A. (2022). Employees' CSR attributions and pro-environmental behaviors in the hotel industry: The key role of female supervisors. *The Service Industries Journal.* https://doi.org/10.1080/02642069.2022.2041604

Wang, J., Wang, S., Xue, H., Wang, Y., & Li, J. (2018). Green image and consumers' word-of-mouth intention in the green hotel industry: The moderating effect of millennials. *Journal of Cleaner Production*, *181*, 426–436.

Yadav, R., Balaji, M. S., & Jebarajakirthy, C. (2019). How psychological and contextual factors contribute to travelers' propensity to choose green hotels? *International Journal of Hospitality Management*, *77*, 385–395.

Yadav, R., Dokania, A. K., & Pathak, G. S. (2016). The influence of green marketing functions in building corporate image: Evidences from hospitality industry in a developing nation. *International Journal of Contemporary Hospitality Management*, *28*(10), 2178–2196.

Part IV

Gen Z and ethical consumption

10 Generation Z lifestyle

Food activism and sustainable traveller behaviour

Alicia Orea-Giner

Introduction

The culinary system is a collective production; it is part of the way of life of a social group. However, different culinary systems can coexist in our societies, so the decision to adopt one food practice, for one product or another, is linked to a personal decision concerning our lifestyle. As consumers, we have the 'freedom' to choose what to eat at any given moment.

Food tourism, culinary tourism, or gastronomic tourism are phrases used to describe how travellers' primary purpose is to visit various features inherent in a tourist destination's food services, such as food outlets, food festivals, and food producers (Hall & Sharples, 2003; Horng & Tsai, 2012; Ellis et al., 2018). Previous studies examined the growth of food tourism research, considering its role in driving food production (Hall & Gössling, 2016) and its implications for sustainability (Yeoman & McMahon-Beatte, 2016; Baum, 2019). Okumus et al. (2018), for example, identified a substantial relationship between sustainability and food tourism (see also Gössling & Hall, 2022).

Okumus (2020) and Hall (2020) emphasise the importance of future food tourism studies on food waste, environmental concerns, and sustainability issues. Previous research has also found a link between lifestyle and customer purchasing (Damijanić, 2019). According to Haenfler et al. (2012), lifestyle encourages social movements associated with social change, and they use the phrase 'lifestyle movements' to refer to individual and communal action to alter society. Indeed, there are certain lifestyle fads that are linked to sustainability, such as vegetarianism, minimalism, and Zero Waste. Taking this method, the purpose of this article is to determine how food tourism experiences are related to sustainable food consumption.

This chapter aims to analyse the influence of Gen Z on tourism and leisure-related decisions through their participation in a lifestyle movement and virtual community on Instagram and TikTok, called 'Realfooding'. Considering these aspects, the following exploratory research question arises:

RQ. Can Gen Z's participation in a lifestyle movement and virtual community influence their tourism and leisure-related decisions?

DOI: 10.4324/9781003289586-14

Lifestyle concept and Gen Z

The definition of lifestyle includes one's way of eating, cooking, and generally nourishing oneself (Featherstone, 1987). It is a term that is located within contemporary consumer culture, indicating individuality, self-expression, and a kind of self-awareness. This concept also includes the way of speaking, clothing, leisure activities, home, choice of holidays, etc., which are indicators of the consumer's tastes and style (Featherstone, 1987). However, these lifestyles also promote collective identities, and a lifestyle can be a vehicle for social movements, in the sense of promoting social change (Haenfler et al., 2012). Haenfler et al. (2012) propose the concept of 'lifestyle movements' to refer to this social phenomenon. They specify that 'lifestyle movements' promote individual and private action to achieve collective change. These types of movements also suggest that the actions of each person lead to social change and create a shared identity, which helps each person find their own identity (Haenfler et al., 2012).

Consumer culture is also linked to the concept of lifestyle movements. Consumer culture allows us to analyse the cultural dimension of the economy as well as the economics of cultural goods (Featherstone, 1987). Haenfler et al. (2012) point out that lifestyles serve to identify oneself and to differentiate oneself from others. However, there are some lifestyles that are related to food that challenge the dominant cultural norms in each cultural context, such as veganism. This is a type of lifestyle that is identified as part of political action in defence of the rights of animals to a dignified life, focused on promoting change at the societal level, although it is initiated at the individual level. In this way, consumption can be understood as linked to a form of activism. This activism is not limited to protest or boycott actions against specific brands or products, but also to the selection of specific consumption options that generate a distinct lifestyle. Due to this fact, lifestyle movements are characterised by their advocacy of lifestyle choice as a tactic for social change, the importance of personal identity work, and the diffuse structure of lifestyle movements. These social lifestyle movements promote individual action by focusing on aspects of daily consumption such as consumption habits, leisure, eating, cooking, and resource consumption, among others (Haenfler et al., 2012).

Lifestyle movements rely on cultural entrepreneurs, social networks, and shared media to shape an ongoing movement discourse and provide some degree of structure (Haenfler et al., 2012). This is an important aspect, as it implies not only a protest action or a change in lifestyle and consumption but also the development of an entrepreneurial spirit and a market that provides the consumer goods demanded by the movement. Therefore, the concept of 'market activism' (Orea Giner, 2021), as distinct from 'consumer activism', is used to refer to the processes by which consumers become producers or entrepreneurs to bring products to the market in line with their lifestyle. Market activism is defined as a series of actions undertaken to promote changes in the production, distribution, and consumption of food products through one's own participation in the market by introducing co-creation into existing products for modification and adaptation, as well as the creation of other products based on the categorisation and cataloguing of the lifestyle foods in question.

In relation to lifestyles, there are different social movements focusing on food. Veganism and vegetarianism are among the lifestyles that have been previously identified in the literature. Cherry (2006) highlights that veganism, like other cultural movements, can be linked to membership in organisations related to the movement but is not compulsory. However, through their participation, their collective identity and their own lifestyle, a type of activism is promoted (Cherry, 2006). Frawley (2013) also suggests that a vegetarian or vegan lifestyle can make it hard to make new friends and keep old ones. This is because the choice of diet can make it hard to make friends with people who don't follow a vegan diet or who don't share this social system.

Food and lifestyle movements have had a significant impact on Gen Z. This generation is a challenge since it appears that they behave differently to earlier generations, which can lead to changes in consumer behaviour (Schlossberg, 2016; Gössling & Hall, 2022). Su et al. (2019) identified that sustainable activists, sustainable believers, and sustainable moderates among this group of individuals placed different importance on their food choices. Gen Z consumers follow healthy eating habits, and their purchasing decisions are linked to sustainable activism (Su et al., 2019).

Food attributes influence consumer choices in various 'green' operational practices considering previous studies (Botonaki et al., 2006; Cavagnaro & Staffieri, 2015; Mamula Nikolić et al., 2021). However, there are scarce studies on sustainable food attributes and traveller behaviour within this generation. Consequently, a more thorough understanding of the relationship between sustainable food attributes and consumer ecological behaviour is necessary (Su et al., 2019).

Food activism, sustainability, and traveller behaviour

Customers' lifestyles have a direct effect on their leisure choices (DaSilva et al., 2020). Consumers' purchase decisions related to tourism and leisure services are affected by wellness and healthy lifestyles (Damijanić, 2019). Lifestyle movements are focused on lifestyle choices to promote social change in one's identity (Haenfler et al., 2012). Some lifestyle movements relate to the concept of 'food activism', which is defined by Siniscalchi and Counihan (2014, p. 3) as "the efforts carried out by people to change the food system around the world by modifying the way of producing, distributing and/or consuming food". In this way, they promote food consumption and transform the food system into one that is more ethical, sustainable, healthy, democratic, culturally appropriate and of higher quality (Siniscalchi & Counihan, 2014). Through the analysis of their dishes, 'food activists' denounce injustices in the food system while promoting changes in eating habits to push for a change in consumption in favour of their notion of 'good food' (Alkon & Guthman, 2017). Schneider et al. (2017) identified different types of forms of food activism using different media since it can be done through web pages, blogs, social networks, or mobile applications. Besides, the narratives and the conversation generated in social networks about food show that consumers play a crucial role in activism to change food practices and the entire food system (Schneider et al., 2017).

The study of the food system and how food is produced, sold, and eaten throughout its local and worldwide history helps to understand a place's cultural past and is linked to tourist activities (Fusté-Forné, 2020). Eating enthusiasts build their identities around gastronomy, ecological consumerism, a healthy approach to food and participating in food-related activities when they travel (Andersson & Mossberg, 2017).

Methodology

This study is focused on conducting qualitative research by participant observation through virtual ethnography. The main methodologies employed for studying online communities are netnography and virtual ethnography. Netnography is defined by Kozinets (2015) as a set of elements that enable data collection, analysis, and ethical research on collected data and participant observation through free data found on the Internet, including mobile applications. However, in this study, we study online practices and the extrapolation of these practices to everyday life, making it more appropriate to conduct a virtual ethnography (Ardévol & Gómez-Cruz, 2014). This method enables the study of both Internet usage and social practices online, as well as how these practices affect individuals (Ardévol & Gómez-Cruz, 2014). As mentioned, the study focuses on the 'Realfooding' lifestyle movement community.

The participant observation was carried out using Instagram and TikTok, considering the public accounts focused on promoting the 'Realfooding' lifestyle movement. Instagram is a prominent tool for influencer marketing in the food and beverage industry, promoting food and beverage firms (Barbe & Neuburger, 2021). A virtual ethnography was conducted from October 1, 2021, to February 12, 2022. Before starting it, the ethnographer created an Instagram and TikTok account, explaining the object of its social media presence. This process includes one of the researchers following the 'Realfooding' lifestyle movement and following the Realfooding criteria of food consumption and travelling considering it. This meant that the ethnographer had to change how they bought and ate food every day.

The second stage of our research is based on narrative research (Bertaux, 2016) based on online interviews and the analysis of the Instagram and/or TikTok accounts of ten individuals (Table 10.1). The participants follow the 'Realfooding' lifestyle movement and were born between 1995 and 2010, following the classification of generations proposed by Williams and Page (2011), to produce an autobiographical narrative of their experience in this virtual community (Rickly-Boyd, 2016). Narratives provide a highly emotional dimension (Bruner, 2010). Telling an individual's experience helps build a symbolic universe and grasp the feelings expressed by the person to experience something unique. Narratives also have the advantage of studying action over time.

Table 10.1 Participant profile

Year of birth	Instagram	TikTok	Profession	Lifestyle	Code
1997	Yes	No	Assistant teacher	Flexitarian Realfooding	P1
1997	Yes	Yes	Teacher	Realfooding	P2
2000	Yes	Yes	Student	Realfooding	P3
1999	Yes	Yes	Nurse	Realfooding	P4
1996	Yes	No	Hospital management assistant	Realfooding	P5
1996	Yes	No	Researcher	Realfooding	P6
1996	Yes	Yes	IT support	Flexitarian. Realfooding	P7
1999	Yes	No	Shop assistant	Vegan. Realfooding	P8
1998	Yes	Yes	Student	Realfooding	P9
1998	Yes	No	Teacher	Realfooding	P10

The online interviews were divided into four topics: (1) lifestyle movements (Featherstone, 1987; Haenfler et al., 2012; Cherry, 2006); (2) social media activism (Grasseni, 2014; Gretzel, 2017; Rousseau, 2013; Saboia et al., 2018); (3) travelling following the 'Realfooding' lifestyle movement (Gracia-Arnaiz, 2015; Schlossberg, 2016; Haddouche & Salomone, 2018; Valencia et al., 2019; Damijanić, 2019); and (4) sustainable choices when travelling (Featherstone, 1987; Schlossberg, 2016).

Results and discussion

The results have been structured into two blocks. The first block comprises the information extracted when conducting the participant observation, obtaining results about the 'Realfooding lifestyle movement' and its connection with food activism and sustainability. The second block is focused on analysing the results from the narrative research, considering the influence of food activism on participants' travel behaviour.

Realfooding lifestyle movement: food activism and sustainability

The 'Realfooding' lifestyle movement and community are linked to 'food activism'. Following the principles of this lifestyle movement can result in changes in consumption habits, criticism of the agri-food industries' power and political ties and is linked to community development strategies. Its activism is mainly deployed on social media. One of the pillars of the 'Realfooding' lifestyle movement is to focus efforts on leading a healthy diet that is extrapolated to lifestyle changes, allowing for greater awareness about product consumption and the influence that marketing has on purchasing decisions. Other trends or related lifestyles are also identified, including veganism, minimalism, and Zero Waste. The current impact of the 'Realfooding' lifestyle movement can be detected by analysing social networks. This community is on Facebook (more than 175,000 people following the movement's page) and Instagram (more than a million followers), even launching its own application for smartphones with its social network included (more than 170,000 users). This community is especially big in Spain, but because the content on social networks is global, people from all over the world, especially from Spanish-speaking countries, can join.

The categorisation and labelling of foods as 'real food' are based on the food itself, that is, not on nutrients or calories. They, therefore, emphasise that to be healthy, food should be as natural as possible and have minimal processing. Therefore, they eat so-called 'real food' and avoid ultra-processed foods. The community clearly defines which foods are included and which are not. The 'Realfooding' lifestyle movement establishes food consumption patterns based on scientific evidence and research on the consumption of ultra-processed foods and has three basic rules. Firstly, food should be based on products that have been minimally processed. Secondly, the diet can be based on so-called 'good processed' products. Finally, it is recommended to avoid consuming ultra-processed products, stressing that their consumption can be occasional, that is, once a month, especially in social settings (Ríos, 2019). No exclusions are made, but it is true that to feel part of the community, it is necessary to comply with a dietary regime based on at least 90% of unprocessed or good processed foods and 10% of ultra-processed foods.

The 'Realfooding' lifestyle movement also promotes the use of scientific studies on nutrition and consumption habits to establish the necessary criteria regarding food and other habits such as the consumption of alcohol, tobacco, and physical exercise. In other words, as mentioned earlier, there is a link between healthy living and the analysis of the results of scientific studies to apply them to lifestyle and thus advance a healthy and fulfilling life. The consumption of 'real food' and 'good processed foods' can also be linked to the concept of 'clean eating', which is a moral system of food consumption, with a preference for foods perceived as natural, free of chemicals, additives, preservatives and refined or ultra-processed ingredients (Allen et al., 2018).

Therefore, 'Realfooding' is a lifestyle movement focused on promoting food selection based on the classification pattern between real food, 'good processed' and 'ultra-processed'. Featherstone (1987) and Haenfler et al. (2012) emphasise that lifestyle is embedded in contemporary consumer culture. Actions that promote social change related to eating habits can be considered lifestyle movements. In this way, the food choices made by 'Realfooding' followers impact the decisions and purchasing habits of food products. Therefore, the 'Realfooding' lifestyle movement would be defined as a social movement based on a healthy lifestyle. It combines the consumption of real food with a life based on healthy habits, which aims to bring about a change in society and its consumption patterns, but, at the same time, to generate a change in the structure of the food and catering market through the individual actions of each person who makes up this community. This movement embraces different actions linked to sustainability, such as buying products in the local market, consuming seasonal products, avoiding the consumption of ultra-processed products associated with high use of pollutants such as plastic and trying to avoid food waste.

From the recommendations or consumption guidelines promoted through Instagram, a reduction in meat consumption is included. It is also a current topic due to the publication of different scientific articles linked to meat consumption, malnutrition, and the effect it has on climate change. Considering the characteristics of the 'Realfooding' lifestyle movement and its symbolic codes, it is also possible to detect that within the community itself, there are divisions and subgroups. Because of this fact, it is possible to follow the 'Realfooding' lifestyle movement and other lifestyle movements such as veganism simultaneously (e.g., the @Realfooding. vegan account). This means that within the 'Realfooding' lifestyle movement, the interrelationships with other lifestyles and political positions such as veganism or vegetarianism are also contemplated. As such, it should be noted that vegetarianism and veganism are social movements related to food and political positioning, which can also be linked to 'Realfooding'. These social movements can promote lifestyle activism (Cherry, 2006), as can 'Realfooding'. Therefore, there is a relationship with other movements, such as veganism and Zero Waste.

Finally, it is important to consider that 'Realfooding' as a healthy lifestyle movement, which presents its own culinary system and encourages individualism in the search for a whole life through the consumption of food and products validated by their classification, as well as other activities, such as sport, is not only practised at an individual level to achieve optimal levels of health. The 'Realfooding' lifestyle movement also shows a form of activism carried out from the body and through consumption and the market in a collective way, that is, by the community. This lifestyle movement is also developing food activism activities by launching products under its own trademark ('Realfooding') that are offered for sale in supermarkets in order, according to the movement's leader, to change the system. It is called 'market activism' (Orea Giner, 2021).

Gen Z and the "Realfooding" lifestyle movement: food activism and sustainable travel behaviour

Gen Z is one of the most active generations in the 'Realfooding' lifestyle movement and the Millennial generation. Through participant observation, it is possible to confirm that Gen Z presents an interest in following a healthy lifestyle while at the same time combining it with ethical consumption (Walters, 2021). Considering that their lifestyle influences consumption, their travel behaviour is also influenced. Gen Z is a powerful force that can influence health and well-being expectations regarding food consumption (Kılıç et al., 2021).

Through the narratives of the participants in this study, it is possible to see how their lifestyles are based on the consumption of healthy foods, although when travelling and leisure activities are involved, it is difficult to make healthy and sustainable choices, as the following interviewee highlights:

I try to look for healthy food options, but I usually always end up failing and consuming more ultra-processed or less healthy foods than I should . . .

Despite staying in a hotel, we made home-made sushi on the first night and purchased healthy food and snacks to snack on between meals rather than eating whatever we could find. Nevertheless, we ate in different restaurants and tried the typical food of the area.

                                                                                                                P6

Gen Z represents a tremendous opportunity for hospitality and tourism firms. Well-being is related to physical, mental health, social, technological/digital, and environmental aspects of the well-being of Gen Z. Hospitality and tourism providers must continue to utilise these dimensions when creating products, services, and experiences (Olson & Ro, 2021). In the narratives, experiences during the trips are mentioned. A search for healthy activities and organising holidays by trying to make healthy choices instead of other options can be detected. One of the interviewees commented on this aspect:

I travelled with my partner, with whom I share this lifestyle, to two different cities. We both try to balance our healthy and unhealthy habits (beer and partying, ultra-processed food, etc.). We had breakfast in the flat, just like at home, toast, coffee, fruit and yoghurt; we walked everywhere to stay active and get to know all the streets of the city. The rest of the meals were usually out of the house, but we didn't worry too much about that, we just ate what we felt like eating.

                                                                                                                P3

However, healthy living is seen as a continuous process, to be considered on a day-to-day basis and occasional consumption of other foods on holiday is not a problem if it allows you to get to know the local gastronomy or to try new products. This fact is also linked with sustainable choices when travelling. Many members of Gen Z will spend more money on immersive and authentic experiences that allow them to live life as a local rather than a typical visitor (Walters, 2021) and consider different lifestyles. It is also related to the concept of food tourism. The global food production system must adjust its offer to lifestyle consumers, considering the trend towards healthy habits and high-quality eating experiences (Orea-Giner & Fusté-Forné, 2022). Through the analysis of the interviews, it is possible to identify two types of travellers: one more concerned about following their lifestyle every day and, also, when travelling; and the other one focused on doing it in their daily life, considering that travelling is a different situation that can allow them to have unusual behaviour. Understanding the tourist's behaviour at home as well as when they travel may assist in identifying possibilities to profit from sustainability marketing to environmentally or socially concerned travellers (Holmes et al., 2021). Some of the participants highlight that their changes are focused on consuming local products and considering avoiding the generation of a large amount of waste. The interest of Gen Z in ethical consumption (Walters, 2021) is connected to food waste reduction initiatives in the tourism industry

(Goh & Jie, 2019). This is confirmed by the following statement from one of the interviewees:

> Buying the most products in bulk, in this way, products packaged with plastic become secondary. Also, by consuming better quality products, you know that their care will always be more sustainable, they will be better cared for, and they will always have a better quality.
>
> P9

Gen Z cares about green, sustainable, organic and local products. These young people are also curious about learning about food using YouTube channels or Instagram (Kılıç et al., 2021). They share content about their lifestyle on social networks. However, they do not consider themselves influencers on social networks, but through their social networks and in their day-to-day lives, they share their lifestyle with other people to change consumption patterns. However, they stress that their generation is more aware of the need to make sustainable decisions and that this is due to the availability of information and easy access to it, as the following interviewee considers:

> More and more people are joining the change and making noise through social media. As well as being the most aware generation, we are also the most influential and I believe that all this will mean a change (P3).
>
> We are more aware of climate change, macro-farms, etc. and we are trying to improve and think about our nutrition.
>
> P8

Barbe and Neuburger (2021) highlight that sustainability and climate change remain important topics for Gen Z, and influencers promoting sustainable travel are increasingly popular on social media platforms. Food waste reduction measures in the tourism sector are also critical for matching a sustainable offer with a sustainable demand (Goh & Jie, 2019). This awareness of sustainability is linked, in the participants' imagination, to healthy living. They consider food, physical exercise, and mental health to be fundamental to healthy living. However, they also include sustainability when talking about the origin of products, the way they are consumed and their quality. They also claim that healthy living leads to a full and happy life, which has an impact on minimising negative impacts on society and the environment.

> It is the future of consumption, more authentic and ethical food for the environment, responsibility for our planet and our own internal bodies.
>
> P9

Conclusions

Understanding environmentally responsible behaviour (Mamula Nikolić et al., 2021) regarding all forms of tourism is becoming increasingly important, including dimensions related to the environment, society, health, or food. Understanding young tourists' behaviour is crucial because they are essential for tourism's future. They have different needs and wants compared to contemporary middle-aged tourists. Thus, to prepare for the future, the tourism sector must understand and cater to these changing needs (Cavagnaro et al., 2018). Gen Z is a powerful force with the potential to influence health and well-being expectations around food intake (Kılıç et al., 2021). Specifically, Gen Z has demonstrated their strong views about the world, their future, and the need to rethink the relationships between people and the planet. They are armed with learning from humanity's previous missteps, environmentally aware and tend to rally spontaneously behind global causes that resonate with them (Bogueva & Marinova, 2020).

Food representations on social media have increased public understanding of the diverse global food production, distribution, and consumption chains (Ranteallo & Andilolo, 2017). Food activism is increasingly supported by online and social media, which offer users different affordances, supporting activism in various ways and leading to unique activist activities. Almost all activism now includes social media activities (Gretzel, 2017). Primarily, activism linked to food consumption is also related to the development of communities (Rousseau, 2013). Thus, by studying the literature focused on food activism, a need arises to analyse how food activism can have economic, political, and social implications and generate interactions between different people (Schneider et al., 2017). The concept of 'market activism' (Orea Giner, 2021) relies on social media to generate an individual level of people who follow lifestyles related to the products being sold to change the market from the market itself. This type of activism is also related to food, with Gen Z actively participating.

Gen Z shows an interest in following a healthy lifestyle and promoting it by following different initiatives that have appeared on social media. The lifestyle that arises from the 'Realfooding' movement is linked to a more sustainable life and considers the scientific literature to make better choices as consumers. Consequently, this lifestyle movement provokes a new form of travel and induces more responsible and sustainable choices, especially for Gen Z consumers.

Suggestions and practical implications

This research provides a starting point for the tourism sector to rethink its services' future and engage in 'market activism'. While other generations have changed their consumption patterns as the market has changed and marketing research has designed products to meet consumer needs or create new ones, Gen Z is considering a change. The ease with which they can consult scientifically contrasted information and their interest in sustainability have led this generation to promote activism in their daily lives linked to the purchasing and consumption decisions they make. The tourism industry is not yet prepared to cope with Gen Z. It is not a matter of adapting the service or carrying out 'brandwashing' strategies. Tourism services, and explicitly catering, must be ethical and responsible with their decisions from the emergence of the brand and the concept of the service. This fact is a fundamental aspect of attracting Gen Z.

References

Alkon, A. H., & Guthman, J. (2017). A new food politics. In A. Alkon & J. Guthman (Eds.), *New food activism: Opposition, cooperation, and collective action* (pp. 316–324). University of California Press.

Allen, M., Dickinson, K. M., & Prichard, I. (2018). The dirt on clean eating: A cross sectional analysis of dietary intake, restrained eating and opinions about clean eating among women. *Nutrients, 10*(9), 1266.

Andersson, T. D., & Mossberg, L. (2017). Travel for the sake of food. *Scandinavian Journal of Hospitality and Tourism, 17*(1), 44–58.

Ardévol, E., & Gómez-Cruz, E. (2014). Digital ethnography and media practices. In A. N. Valdivia (Ed.), *The international encyclopedia of media studies* (Vol. 7, pp. 498–518). John Wiley and Sons Ltd.

Barbe, D., & Neuburger, L. (2021). Generation Z and digital influencers in the tourism industry. In N. Stylos, R. Rahimi, B. Okumus, & S. Williams (Eds.), *Generation Z marketing and management in tourism and hospitality* (pp. 303–325). Palgrave Macmillan.

Baum, T. (2019). Bridging the gap: Making research "useful" in food, tourism, hospitality and events-the role of research impact. In S. Beeton & A. Morrison (Eds.), *The study of food, tourism, hospitality and events* (pp. 157–166). Springer.

Bertaux, D. (2016). *Le récit de vie*. Armand Colin.

Bogueva, D., & Marinova, D. (2020). Cultured meat and Australia's generation Z. *Frontiers in Nutrition, 7*, 148.

Botonaki, A., Polymeros, K., Tsakiridou, E., & Mattas, K. (2006). The role of food quality certification on consumers' food choices. *British Food Journal, 108*(2), 77–90.

Bruner, J. S. (2010). *Pourquoi nous racontons-nous des histoires?: Le récit, au fondement de la culture et de l'identité*. Retz.

Cavagnaro, E., & Staffieri, S. (2015). A study of students' travellers values and needs in order to establish futures patterns and insights. *Journal of Tourism Futures, 1*(2), 94–107.

Cavagnaro, E., Staffieri, S., & Postma, A. (2018). Understanding millennials' tourism experience: Values and meaning to travel as a key for identifying target clusters for youth (sustainable) tourism. *Journal of Tourism Futures, 4*(1), 31–42.

Cherry, E. (2006). Veganism as a cultural movement: A relational approach. *Social Movement Studies, 5*(2), 155–170.

Damijanić, A. T. (2019). Wellness and healthy lifestyle in tourism settings. *Tourism Review, 74*(4), 978–989.

DaSilva, G., Hecquet, J., & King, K. (2020). Exploring veganism through serious leisure and liquid modernity. *Annals of Leisure Research, 23*(5), 627–644.

Ellis, A., Park, E., Kim, S., & Yeoman, I. (2018). What is food tourism? *Tourism Management, 68*, 250–263.

Featherstone, M. (1987). Lifestyle and consumer culture. *Theory, Culture & Society, 4*(1), 55–70.

Frawley, W. (2013). *Linguistic semantics*. Routledge.

Fusté-Forné, F. (2020). Say Gouda, say cheese: Travel narratives of a food identity. *International Journal of Gastronomy and Food Science, 22*, 100252.

Goh, E., & Jie, F. (2019). To waste or not to waste: Exploring motivational factors of Generation Z hospitality employees towards food wastage in the hospitality industry. *International Journal of Hospitality Management, 80*, 126–135.

Gössling, S., & Hall, C. M. (2022). *The sustainable chef: The environment in culinary arts, restaurants, and hospitality.* Routledge.

Gracia-Arnaiz, M. (2015). *Comemos lo que somos.* Icaria Editorial.

Grasseni, C. (2014). Food activism in Italy as an anthropology of direct democracy. *Anthropological Journal of European Cultures, 23*(1), 77–98.

Gretzel, U. (2017). Social media activism in tourism. *Journal of Hospitality and Tourism, 15*(2), 1–14.

Haddouche, H., & Salomone, C. (2018). Generation Z and the tourist experience: Tourist stories and use of social networks. *Journal of Tourism Futures, 4*(1), 69–79.

Haenfler, R., Johnson, B., & Jones, E. (2012). Lifestyle movements: Exploring the intersection of lifestyle and social movements. *Social Movement Studies, 11*(1), 1–20.

Hall, C. M. (2020). Improving the recipe for culinary and food tourism? The need for a new menu. *Tourism Recreation Research, 45*(2), 284–287.

Hall, C. M., & Gössling, S. (Eds.). (2016). *Food tourism and regional development: Networks, products and trajectories.* Routledge.

Hall, C. M., & Sharples, L. (2003). The consumption of experiences or the experience of consumption? An introduction to the tourism of taste. In C. M. Hall, L. Sharples, R. Mitchell, N. Macionis, & B. Cambourne (Eds.), *Food tourism around the world* (pp. 1–24). Routledge.

Holmes, M. R., Dodds, R., & Frochot, I. (2021). At home or abroad, does our behavior change? Examining how everyday behavior influences sustainable travel behavior and tourist clusters. *Journal of Travel Research, 60*(1), 102–116.

Horng, J. S., & Tsai, C. T. (2012). Culinary tourism strategic development: An Asia-Pacific perspective. *International Journal of Tourism Research, 14*(1), 40–55.

Kılıç, B., Bekar, A., & Yozukmaz, N. (2021). The new foodie generation: Gen Z. In N. Stylos, R. Rahimi, B. Okumus, & S. Williams (Eds.), *Generation Z marketing and management in tourism and hospitality* (pp. 223–247). Palgrave Macmillan.

Kozinets, R. V. (2015). *Netnography: Redefined.* Sage.

Mamula Nikolić, T., Pantić, S. P., Paunović, I., & Filipović, S. (2021). Sustainable travel decision-making of Europeans: Insights from a household survey. *Sustainability, 13*(4), 1960.

Okumus, B. (2020). Food tourism research: A perspective article. *Tourism Review, 76*(1), 38–42.

Okumus, B., Koseoglu, M. A., & Ma, F. (2018). Food and gastronomy research in tourism and hospitality: A bibliometric analysis. *International Journal of Hospitality Management, 73*, 64–74.

Olson, E. D., & Ro, H. (2021). Generation Z and their perceptions of well-being in tourism. In N. Stylos, R. Rahimi, B. Okumus, & S. Williams (Eds.), *Generation Z marketing and management in tourism and hospitality* (pp. 101–118). Palgrave Macmillan.

Orea Giner, A. (2021). *Alimentación, estilo de vida y activismo de mercado: Una etnografía virtual en Instagram sobre el "realfooding".* Universitat Oberta de Catalunya. http://hdl.handle.net/10609/134033

Orea-Giner, A., & Fusté-Forné, F. (2022). Beyond fueling our bodies to feeding our minds. *Journal of Sustainability and Resilience, 2*(1). https://digitalcommons.usf.edu/jsr/vol2/iss1/1

Ranteallo, I. C., & Andilolo, I. R. (2017). Food representation and media: Experiencing culinary tourism through foodgasm and foodporn. In A. Saufi, I. Andilolo, N. Othman, & A. Lew (Eds.), *Balancing development and sustainability in tourism destinations* (pp. 117–127). Springer.

Rickly-Boyd, J. M. (2016). Dirtbags: Mobility, community and rock climbing as performative of identity. In T. Duncan, S. A. Cohen, & M. Thulemark (Eds.), *Lifestyle mobilities: Intersections of travel, leisure and migration* (pp. 51–64). Ashgate Publishers.

Ríos, C. (2019). *Come comida real*. Editorial Paidós.

Rousseau, S. (2013). *Food media: Celebrity chefs and the politics of everyday interference*. Berg.

Saboia, I., Almeida, A. M. P., Sousa, P., & Pernencar, C. (2018). I am with you: A netnographic analysis of the Instagram opinion leaders on eating behavior change. *Procedia Computer Science*, *138*, 97–104.

Schlossberg, M. (2016). Teen Generation Z is being called "millennials on steroids", and that could be terrifying for retailers. *Business Insider*. www.businessinsider.com/millennials-vs-gen-z-2016-2?r=US&IR=T

Schneider, T., Eli, K., Dolan, C., & Ulijaszek, S. (Eds.). (2017). *Digital food activism*. Routledge.

Siniscalchi, V., & Counihan, C. (2014). Ethnography of food activism. In C. Counihan & V. Siniscalchi (Eds.), *Food activism: Agency, democracy and economy* (pp. 3–14). Bloomsbury.

Su, C. H., Tsai, C. H., Chen, M. H., & Lv, W. Q. (2019). US sustainable food market Generation Z consumer segments. *Sustainability*, *11*(13), 3607.

Valencia, E. L., López, I. G., & Patiño, A. C. (2019). El espacio culinario: Una propuesta de análisis desde la Antropología de la Alimentación. *Antropología Experimental*, *19*(15), 165–172.

Walters, P. (2021). Are Generation Z ethical consumers? In N. Stylos, R. Rahimi, B. Okumus, & S. Williams (Eds.), *Generation Z marketing and management in tourism and hospitality* (pp. 303–325). Palgrave Macmillan.

Williams, K. C., & Page, R. A. (2011). Marketing to the generations. *Journal of Behavioral Studies in Business*, *3*(1), 37–53.

Yeoman, I., & McMahon-Beatte, U. (2016). The future of food tourism. *Journal of Tourism Futures*, *2*(1), 95–98.

11 Gen Z tourists' perceptions of ethical consumption

A developing country perspective

Abolfazl Siyamiyan Gorji, Seyedasaad Hosseini, Fernando Almeida Garcia, and Rafael Cortes Macias

Introduction

A more sustainable future can be achieved by consumers' consumption decisions. Words such as sustainable, socially friendly, ethical, environmentally friendly, and responsible are frequently used to represent the link between ethical consumption and tourist behaviour. There is an increasing awareness among consumers about the impact of their consumption on the environment and society in general (Berki-Kiss & Menrad, 2022). According to Harrison et al. (2005), ethical consumers have a variety of reasons for choosing a particular product over another, such as political, religious, spiritual, environmental, social, or other motives. The concept of 'ethical consumption' refers to a consumption behaviour that adheres to individual and moral beliefs while practising social responsibility (Chi, 2022). Prior studies have found that there exist ethically minded tourists such as Generation Z (Gen Z) tourists (Seyfi et al., 2022).

Researchers have found that Millennials and Gen Z consumers are not just searching for products and services that offer a good price, quality, brand image, status, or fashion trend when they make purchases, but they're also looking for ethical goods and services and are also willing to pay more for ethical products (Akintimehin et al., 2022). Due to consumers' ethical awareness when defining their shopping strategies, ethical consumption has received much attention from researchers and policymakers (Le et al., 2020).

Most of the previous studies have examined tourists' attitudes towards eco-friendly and green products in industrialised societies. Despite this, very little research has examined the behaviour of 'Zoomers' in developing countries, particularly Iran, regarding sustainable consumption. So, it can be noted that there is a clear gap in the literature between Gen Z consumption patterns and ethical consumerism. Therefore, the aim of this research is to gain a better understanding of Iranian Gen Z's awareness of sustainability, their attitude towards sustainable products, and their willingness to pay for them. For consumer behaviour researchers, the question of what drives Gen Z to choose ethical products remains a puzzle (Djafarova & Foots, 2022). The true motivations behind Gen Z's ethical purchase consumption are still not well studied in marketing and tourism research. Thus, this study adopted in-depth interviews with the Iranian Gen Z to address

DOI: 10.4324/9781003289586-15

the research objectives: (1) understanding the level of sustainability consciousness and ecological awareness of the Iranian Gen Z; (2) identifying Iranian Gen
Z tourists' attitudes towards ethical and environmental concerns today; and (3)
exploring the role of subject norms and perceived behavioural controls on their
ethical consumption.

Literature review

Gen Z and Z tourists

A new generation of human beings is known as Gen Z. They are also referred to
as 'Gen Zers' and 'Zoomers' (Khalil et al., 2021). Although Generation Y (Gen
Y), also known as 'the Millennial Generation', has been extensively studied,
Gen Z remains largely unexplored (Haddouche & Salomone, 2018). According
to Ladwein and Ouvry (2007), the notion of generation has several components:
sociological, demographical, historical, genealogical, and familial. Based on
the sociological approach, individuals' values are shaped by historical events,
as well as their political, economic, and cultural contexts (Ward, 1974; Holbrook & Schindler, 1989). Gen Z can, thus, be defined as those individuals who
were born after 1995 and before 2010 (Monaco, 2018; Veluchamy et al., 2016).
Compared to their predecessors, this generation is quite different. Besides having different abilities and social concerns, they also have different motivations
and learning styles (Mavragani & Dionysios, 2022). Gen Z can also be characterised as a very mixed group, community-oriented, more individualistic and
independent, realistic, practical and materialistic, and digital natives (Seyfi
et al., 2022). A key characteristic of Gen Z is its concern for education and
equality, as well as its concern about sustainability. This generation is indeed
realistic and persistent and is much more than a consumer group (Haddouche &
Salomone, 2018).

In a tourism context, it is predicted that Gen Z will be tourism's main target market in the near future. They are also known as smart tourists as they are
accustomed to virtual worlds and very comfortable with them (Djafarova & Foots,
2022). Tourism research has widely applied the generation cohort theory to get
more insight into this generation (Seyfi et al., 2022). Based on this theory, studies have shown that certain generation cohorts prefer different products, vacation
destinations, and activities (Karakaş et al., 2022). According to some scholars,
there are eight main purposes behind travelling by this generation (Moisˇa, 2010);
To experience new and novel things, cultural exchange, volunteering and working
in a new place, learning a new language, education in another country, participating in adventure events, visiting friends and relatives, and leisure tourism. Gen
Z tourist characteristics can also be categorised as less loyal, using more online
information, being influenced by electronic word-of-mouth (eWOM), travelling
with their friends and family, having a high purchase power, and authenticity seeking (Seyfi et al., 2022).

Gen Z and ethical consumerism

The term 'ethical consumption' is employed in this study to cover a range of terms circulating in the literature, including green purchase, green consumerism, ethical behaviour, political consumption, and responsible and environmentally sustainable consumption. From a theoretical perspective, these are complementary (Davies & Gutsche, 2016). Consumer behaviour that takes into account societal norms while influenced by ethical criteria can be termed ethical consumerism (Chatzidakis et al., 2004). According to Cooper-Martin and Holbrook (1993), consumers' ethical behaviour is determined by their ethical concerns in making decisions, buying products, and in general participating in consumption activities. Indeed, ethical consumerism is a type of political action that pertains to the idea that market customers consume indirectly both products and processes (Coffin & Egan-Wyer, 2022). Moreover, ethical consumerism views consumption as a political act that sanctions or legitimises the values and ideals embedded in a product's production (Hassan et al., 2022). Berki-Kiss and Menrad (2022) state that consumers can become more involved in sustainability by engaging in ethical consumption.

In addition, studies have shown that visitors are typically unwilling to embrace sustainable consumption behaviours and do not believe it is their duty to do so (Miller et al., 2010), whilst even those who generally claim to employ ethical consumption behaviour at home do not do so whilst on vacation (Juvan & Dolnicar, 2014). It seems that there could be a shift towards more sustainable tourism consumption in the younger post-Millennial generation ('Gen Z'; Sharpley, 2021). Several surveys have found that the social and environmental effects of their consumption and the need to address global warming are the most important concerns to post-Millennials. In particular, they demonstrate strong environmental and ethical values when purchasing products (Noor et al., 2017). According to Tallontire et al. (2001), ethical consumerism can be positive, negative, or consumer-driven. Micheletti et al. (2008) have subsequently dubbed the later form 'discursive'. Choosing eco-efficient, organic and green products or services is an example of a positive type. The negative form refers to the boycotting of specific items, companies, or groups of companies. In the discursive type, a communication channel is a means of interacting with consumers, forming public opinion through social debate, as well as transforming cultural activities using mainly computers and networks. Walters (2021), suggests five types of ethical consumption; anti-consumerism/sustainable consumerism, relationship purchasing, boycotts, fully screened, and positive buying. Positive ethical action is more likely to be supported by Gen Z (Djafarova & Foots, 2022). Khalil et al. (2021) identified four constructs that drive customers' sustainable consumption, meeting one's needs with high-quality goods without harming environmental health, namely awareness, motivation, action, and advocacy. A study by Stern et al. (1993) identified three types of personal values relevant to ethical consumption: altruistic, biospheric, and egoistic values.

Method

This study sought to ascertain Gen Z tourists' perspectives on ethical consumption in the tourism industry. The authors investigated tourist motivation, attitude, and purchasing behaviour in its entirety in order to gain a better understanding of Gen Z tourists' depth of perspectives. To do so, a semi-structured in-depth interview approach was deemed the most appropriate means for data collection. In this approach, participants' reactions can be observed and their words can be heard accurately (Hosseini et al., 2022). The interview questions were developed based on the literature review and previous studies (e.g., Djafarova & Foots, 2022; Khalil et al., 2021) and were also modified to the study context. The lead author conducted all 20 interviews in Persian which lasted between 30 and 50 minutes. Then, a copy of the manuscript was sent to the respondents in order to check for any words or expressions that might be miscoded. These steps, therefore, ensured the credibility of the interviews (Siyamiyan et al., 2021, 2022a). Researchers adopted purposeful sampling in selecting the participants to be interviewed. In this study, the approach of 'data saturation' was also followed, which refers to no more information observed during additional interviews (Siyamiyan et al., 2022b). All of the participants were junior students in Iran at the time of the study. An initial pilot study with five interviewees allowed the interview questions, interview style, and interview technique to be refined (Kim, 2011). Data analysis in this study was conducted using thematic analysis which is known as a systemic method, flexible, and an increasingly popular method for analysing qualitative data (Hosseini et al., 2022).

Findings and discussion

Attitude and awareness towards sustainability

According to the participant's responses, there is a clear awareness and understanding of the concept of sustainability issues. Moreover, almost all respondents identified themselves as ethically conscious tourists. For instance, one of the participants commented:

> *Most people think that sustainable tourism means collecting garbage at the destination, but the true meaning is helping locals to have a better life through tourism.*
>
> (interview # 3)

> *During travel, not only I try to be ethical, but also try to be responsible as much as I can toward my environment.*
>
> (interview # 1)

Social responsibility is a very important concern for Gen Z members. They understand that their decisions during travel may have a negative impact on others and the environment or compromise access to resources for future generations. It was echoed by one respondent:

> *As a tourist, we must be responsible about the destination where we go, all of us must be worried about our environment and ecosystem.*
>
> (interview #4)

Their attitudes towards environmental and ethical issues are strong, and they are conscious of their consumption. Their awareness of sustainability issues comes mainly from the subjects of their lessons at university or from being exposed to social media:

> *In my lesson at university, I've studied about sustainable concepts and products.*
>
> (interview #9)

> *I gain information mainly from one environmental activist on Instagram. Other than that, there is information about global environmental concerns, such as climate change, destroying forests, and so on, at university subject.*
>
> (interview #8)

According to studies, many consumers improve their awareness based on the information they received through social media (Siyamiyan et al., 2021). Seyfi et al. (2021, 2022) have argued that social platforms can affect consumers' purchase decisions. Indeed, people check social networking sites before purchasing products to see what other people think about them. Social media users have trust in the suggestions and feedback of their colleagues, relatives, and even strangers on these networks.

In addition, they also commented about eco-labelled products. Some participants stated that there is a clear shortage of green destinations as well as green souvenir products in Iran, and they don't have access to eco-souvenirs:

> *I want to buy green souvenirs, but I don't know any eco-shops where we can buy things from our destination. It would be much better if I could find an eco-labeled product.*
>
> (interview #19)

The thematic analyses reveal deep knowledge or awareness of Gen Z about sustainable products and eco-labelled stamps in the tourism industry. They declared that they are well familiar with the issue. However, a few participants expressed that they never saw sustainability stamps on any products related to tourism.

Social pressures and social norms

Another theme that was found was the concerns of the respondents towards the judgements of their peers. The majority of participants expressed concern about how others perceive their travel activities. Indeed, the theory of planned behaviours (Ajzen, 1991) well supports the effects of social norms on individual behavioural intentions. Based on this theory, people's behaviour is influenced by the opinions of other individuals, including friends, family, and colleagues. Indeed, individuals avoid engaging in acts which would be disapproved by the standard of their reference groups (Ajzen & Fishbein, 1975). Normative influence leads to a substantial concern with public presence and an attempt to conform to social expectations (Siyamiyan et al., 2022a). It was echoed by interview number 14 who noted that:

> *If I do mistakes while I'm travelling to a destination, my friends may notice that. I don't want them to think that I do not care about my activities in a destination.*

(interview #14)

Other issues that were found were social image and ethical attitudes. Some of our respondents have stated that if they were worried about their social image, they tend to purchase ethically:

> *Can anyone say that I'm not worried about my social image? Of course, I am worried about what my friends and people around me think about my attitude towards purchasing behaviour or where and how I'm travelling.*

(interview #9)

Djafarova and Foots (2022) have found that social image is one of the key factors in the ethical consumption of Gen Z. They contended that self-presentation and Gen Z tendencies to be seen positively play an important role in shaping their attitude towards engaging in sustainable behaviour. Potentially, even those customers that have no awareness of sustainability and green consumption tend to claim consciousness about these concepts.

Social pressure also motivated a few of the respondents to learn more about the concept of sustainable, ethical consumption. Respondents expressed that the deeper they understand sustainable development, the more committed they are towards their destinations. They also emphasised that green behaviour induces self-appreciation and value. This was echoed by respondents:

> *My friends pushed me to study a bit for these kinds of concepts. So, I looked up things like "ethical consumption", "green products", etc. on Google to keep up with the latest trends in these areas.*

(interview #20)

> *Once I was with my friends at university and they were talking about these kinds of subjects, but I was quiet because I didn't have any idea about the topic. So, later on, I started to study about these notions.*

(interview #3)

Ethical purchase actions and sustainable travel style

Since willingness to purchase green products in tourism is a significant factor of future ethical buying behaviour, respondents were asked "when they want to purchase in a tourism destination, what factors make them purchase or discard a product?". In addition, we asked if they purchase in a manner that could be helpful for the local communities. Some participants stated that they avoid buying non-locally made souvenirs at their destination.

> *I try to buy local things from a destination that I often visit. Sometimes I see imported souvenirs and products, but I avoid buying them.*
>
> (interview #12)

Some of those interviewed expressed their interest to change their lifestyle and their purchasing behaviour in line with sustainable concepts. According to Walters (2021), Gen Z is most interested in integrating sustainability into their travel styles. It was noted by one of the respondents:

> *I've heard that in some international airlines when consumers want to buy a ticket to go to their destination, there is an option to pay more in respect to reduce the CO_2 affects.*
>
> *(green travelling), I do really buy* (interview #18)

This was echoed by another respondent:

> *If I find eco-friendly tour companies in Iran, I will absolutely travel with them.*
>
> (interview #12)

Other participants noted how people can take sustainable travel easily by commenting:

> *Adopting sustainable/responsible travel is not something difficult, just people must know about some small tips. For example, tourists should not print their boarding pass, hotel reservations etc. because they have smart phones and they can save their travel documents on it. These small steps will save a huge amount of paper waste.*
>
> (interview #17)

> *I feel good since I decided to sustainable travel. It is really enjoyable when you travel in way that you caring for the planet.*
>
> (interview #1)

These perspectives could be connected to self-actualisation and self-presentation concepts (Djafarova & Foots, 2022). Davies and Gutsche (2016) refer to the self-satisfaction concept when they argue about feelings with respect to ethical consumption.

Also, some participants declared their inclination to eco-buycott or eco-buy. This concept refers to buying eco-friendly products over other alternatives (Le et al., 2022). Nguyen et al. (2018) argued that since ecologically unfriendly purchase decisions seem to have negative long-term effects, consumers have increasingly shown interest in boycotting products and brands that are detrimental to the environment. Accordingly, environmental guidelines have gained a lot of traction with consumers due to their concern about the environmental impact of disposing of products after consumption and purchase (Göçer & Oflaç, 2017).

Boycott action

The analysis of interviews revealed that 'political consumerism' plays a key role in the initial travel activities decision. Some studies show that boycotting a holiday destination, a tourism-related company or products, or even an accommodation is 'political consumerism' or 'responsible behaviour' (Bostrom et al., 2019; Boulianne, 2022; Siyamiyan et al., 2022a). Lovelock (2012) states that tourists boycott a tourism-related product to show their ethical concerns. Kam and Deichert (2020) affirm that there is a direct impact of collective buycotts and boycotts on public stance, company policies, or policy changes In this respect, Green Match (2020) declare that Gen Z is more willing than previous generations to avoid or refuse to purchase from a company that has not met their standards, leading to boycotting them. The vast majority of respondents in this study emphasised their tendency to participate in a particular campaign or issues which are related to boycotting action or political consumerism. For example:

> *If I understand that the souvenir which I want to buy is not a green product, I will refuse it. I prefer not to buy rather than buying stuff which is harmful for the environment.*
>
> (interview #6)

This was echoed by another respondent:

> *There are some destinations in Iran where some vendors sell the colourful soil of that region. However, I visited there, but I didn't buy, and later I found that there is a boycotting petition on the internet which asks tourists to not buy this soil as a souvenir.*
>
> (interview #8)

In line with consumer sustainability behaviour literature, our findings strengthen the observation that Gen Z tourists are more willing to engage in political consumerism. Prior studies have shown that Gen Z who possess technology skills are more likely to participate in political consumerism and especially ethical consumption

(Djafarova & Foots, 2022). In the case of ecological boycotts, consumers are offended when they perceive a product as damaging to the environment or abusive in some way (Nguyen et al., 2018). Some studies also have demonstrated that since Gen Z are born in the digital era, they are influenced by this space, so that digital media can shape their attitude towards boycott or buycott campaigns (Seyfi et al., 2022).

Conclusion, implications, and limitations

Ethical consumerism and responsible behaviour have become two of the most controversial issues in our lives today due to the global challenges that the world is experiencing, which have impacted tourist travel styles. As a result of these changes, most of the world's young tourists, so-called Gen Z, have become interested in engaging in sustainable consumerism movements. Considering this topic, the current research sought to understand the attitude and motivation of Gen Z tourists towards ethical concepts such as responsible travel, tourism and sustainability, sustainable travel, green travel, ethical consumption, and eco-products. We have considered Iran's Gen Z as an emerging market in tourism in the Middle East region since surprisingly little is documented about ethical consumerism within developing countries (Seyfi et al., 2022). Emerging markets may not equally benefit from consumer theories that emerged in developed markets. As Saarinen (2021) argues, this may be because ethical consumerism is heavily influenced by neoliberal forms of tourist consumption. As a result of understanding such differences, ethical consumerism can be explained in a more comprehensive manner.

In this study, we consider an ethical potential tourist from Gen Z as a person who may have political, social, environmental, or spiritual motives (Harrison et al., 2005; Akintimehin et al., 2022) for choosing or not choosing a tourism-related product, such as souvenirs, accommodation, travel agencies, travel vehicles, and destinations. The study revealed that the Iranians' Gen Zers are conscious of contemporary ethical and environmental challenges. In addition to their awareness that their decisions may affect future generations, Gen Z are aware of the impact their decisions may have on the environment and tourism resources. They are also concerned about their social image when they want to share their opinions on social media about responsible behaviour. Moreover, influencers on social media and recommendations from friends play a big role in enabling Gen Z to take ethical actions during travel.

In addition, our participants perceived limitations in accessing eco-labelled products in the Iranian market. They expressed their interest in ethical purchasing in the case of finding these types of products in holiday destinations. Moreover, we found that what drives Iranian Gen Z consumers to buy ethically was more about their own motivations than a broad generational theme. According to study findings, feeling good is an important motivating factor when Gen Z tourists take ethical action while travelling. In addition, our respondents expressed their tendency to engage in political consumption. They have a tendency to reject destinations, accommodation, and other products which may be harmful to the environment.

This qualitative study provides valuable implications for both scholars and destination marketers. By providing a comprehensive guideline to understanding the Iranian Gen Z with their attitude and motivation to act responsibly in decision-making processes as well as activities in destinations, the study contributes to the existing literature on tourist ethical consumption, since tourism scholars have not sufficiently evaluated these concepts. Using public transportation and buying organic products or eco-brands exemplify the Gen Zers' holistic approach to sustainability and illustrate the variety of strategies destinations can employ to meet their expectations in this area.

This study also has several implications for tourism-related companies in terms of adopting their strategies in line with ethical issues in the tourism industry. Tourism businesses can benefit from the findings emerging from the current study to design some ethical campaigns to attract Gen Z tourists. New startups and businesses can also consider the results of this study to design creative/ethical experiences for their consumers in the tourism sector.

Future research should employ mixed methods to better understand Gen Z tourists' attitudes towards green travel. Moreover, future studies can adopt a cross-cultural approach to evaluate the differences in Gen Z attitudes towards ethical consumption.

References

Ajzen, I. (1991). The theory of planned behavior. *Organizational Behavior and Human Decision Processes*, *50*(2), 179–211.

Ajzen, I., & Fishbein, M. (1975). A Bayesian analysis of attribution processes. *Psychological Bulletin*, *82*(2), 261–277.

Akintimehin, O., Phau, I., Ogbechie, R., & Oniku, A. (2022). Investigating boycotts and buycotts as antecedents towards attitude and intention to engage in ethical consumption. *International Journal of Ethics and Systems*. https://doi.org/10.1108/IJOES-06-2021-0117

Berki-Kiss, D., & Menrad, K. (2022). Ethical consumption: Influencing factors of consumer's intention to purchase fairtrade roses. *Cleaner and Circular Bioeconomy*, *2*, 100008.

Bostrom, M., Micheletti, M., & Oosterveer, P. (Eds.). (2019). *The Oxford handbook of political consumerism*. Oxford University Press.

Boulianne, S. (2022). Socially mediated political consumerism. *Information, Communication & Society*, *25*(5), 609–617.

Chatzidakis, A., Hibbert, S., Mittusis, D., & Smith, A. (2004). Virtue in consumption? *Journal of Marketing Management*, *20*(5–6), 526–543.

Chi, N. T. K. (2022). Ethical consumption behavior towards eco-friendly plastic products: Implication for cleaner production. *Cleaner and Responsible Consumption*, *5*, 100055.

Coffin, J., & Egan-Wyer, C. (2022). The ethical consumption cap and mean market morality. *Marketing Theory*, *22*(1), 105–123.

Cooper-Martin, E., & Holbrook, M. B. (1993). Ethical consumption experiences and ethical space. *Advances in Consumer Research Volume*, *20*, 113–118.

Davies, I. A., & Gutsche, S. (2016). Consumer motivations for mainstream ethical consumption. *European Journal of Marketing*, *50*(7–8), 1326–1347.

Djafarova, E., & Foots, S. (2022). Exploring ethical consumption of Generation Z: Theory of planned behaviour. *Young Consumers*, *23*(3), 413–431.

Göçer, A., & Oflaç, B. S. (2017). Understanding young consumers' tendencies regarding eco-labelled products. *Asia Pacific Journal of Marketing and Logistics*, *29*(1), 80–97.

Green Match. (2020). *4 sustainable behaviours of Gen Z's shopping habits*. www.green match.co.uk/blog/2018/09/gen-zs-sustainable-shopping-habits

Haddouche, H., & Salomone, C. (2018). Generation Z and the tourist experience: Tourist stories and use of social networks. *Journal of Tourism Futures*, *4*(1), 69–79.

Harrison, R., Shaw, D., & Newholm, T. (2005). *The ethical consumer*. Sage Publications.

Hassan, S. M., Rahman, Z., & Paul, J. (2022). Consumer ethics: A review and research agenda. *Psychology & Marketing*, *39*(1), 111–130.

Holbrook, M. B., & Schindler, R. M. (1989). Some exploratory findings on the development of musical tastes. *Journal of Consumer Research*, *16*(1), 119–124.

Hosseini, S., Macias, R. C., & Garcia, F. A. (2022). The exploration of Iranian solo female travellers' experiences. *International Journal of Tourism Research*, *24*(2), 256–269.

Juvan, E., & Dolnicar, S. (2014). The attitude – behaviour gap in sustainable tourism. *Annals of Tourism Research*, *48*, 76–95.

Kam, C. D., & Deichert, M. (2020). Boycotting, buycotting, and the psychology of political consumerism. *The Journal of Politics*, *82*(1), 72–88.

Karakaş, H., Çizel, B., Selçuk, O., Coşkun Öksüz, F., & Ceylan, D. (2022). Country and destination image perception of mass tourists: Generation comparison. *Anatolia*, *33*(1), 104–115.

Khalil, S., Ismail, A., & Ghalwash, S. (2021). The rise of sustainable consumerism: Evidence from the Egyptian generation Z. *Sustainability*, *13*(24), 13804.

Kim, Y. (2011). The pilot study in qualitative inquiry: Identifying issues and learning lessons for culturally competent research. *Qualitative Social Work*, *10*(2), 190–206.

Ladwein, R., & Ouvry, M. (2007). *Entre recherche et production d'expérience dans les environnements commerçants: L'expérience vécue* (No. hal-00199096). https://ideas.repec. org/d/laborfr.html

Le, T. D., Duc Tran, H., & Hoang, T. Q. H. (2022). Ethically minded consumer behavior of Generation Z in Vietnam: The impact of socialization agents and environmental concern. *Cogent Business & Management*, *9*(1), 2102124.

Le, T. D., Nguyen, P. N. D., & Kieu, T. A. (2020). Ethical consumption in Vietnam: An analysis of generational cohorts and gender. *Journal of Distribution Science*, *18*(7), 37–48.

Lovelock, B. (2012). Human rights and human travel? Modeling global travel patterns under an ethical tourism regime. *Tourism Review International*, *16*(3), 183–202.

Mavragani, E., & Dionysios, P. (2022). Gen Z and tourism destination: A tourism perspective of augmented reality gaming technology. *International Journal of Innovation and Technology Management*, *19*(5), 1–15.

Micheletti, M., Follesdal, A., & Stolle, D. (2008). Politics, products, and markets: Exploring political consumerism past and present. *Economic Geography*, *84*(1), 123–125.

Miller, G., Rathouse, K., Scarles, C., Holmes, K., & Tribe, J. (2010). Public understanding of sustainable tourism. *Annals of Tourism Research*, *37*(3), 627–645.

Moisʼa, C. O. (2010). Aspects of youth travel demand. *Annales Universtatis Aplensis Series Oeconomica*, *12*(2), 575–582.

Monaco, S. (2018). Tourism and the new generations: Emerging trends and social implications in Italy. *Journal of Tourism Features*, *4*(1), 7–15.

Nguyen, T. H., Ngo, H. Q., Ngo, P. N. N., & Kang, G. D. (2018). Understanding the motivations influencing ecological boycott participation: An exploratory study in Viet Nam. *Sustainability*, *10*(12), 4786.

Noor, M. N. M., Jumain, R. S. A., Yusof, A., Ahmat, M. A. H., & Kamaruzaman, I. F. (2017). Determinants of Generation Z green purchase decision: A SEM-PLS approach. *International Journal of Advanced and Applied Sciences*, *4*(11), 143–147.

Saarinen, J. (2021). Is being responsible sustainable in tourism? Connections and critical differences. *Sustainability*, *13*(12), 6599.

Seyfi, S., Hall, C. M., Saarinen, J., & Vo-Thanh, T. (2021). Understanding drivers and barriers affecting tourists' engagement in digitally mediated pro-sustainability boycotts. *Journal of Sustainable Tourism*. https://doi.org/10.1080/09669582.2021.2013489

Seyfi, S., Hall, C. M., Vo-Thanh, T., & Zaman, M. (2022). How does digital media engagement influence sustainability-driven political consumerism among Gen Z tourists? *Journal of Sustainable Tourism*. https://doi.org/10.1080/09669582.2022.2112588

Sharpley, R. (2021). On the need for sustainable tourism consumption. *Tourist Studies*, *21*(1), 96–107.

Siyamiyan, A. S., Almeida-García, F., & Mercadé Melé, P. (2021). Analysis of the projected image of tourism destinations on photographs: The case of Iran on Instagram. *Anatolia*. https://doi.org/10.1080/13032917.2021.2001665

Siyamiyan, A., Almeida-García, F., & Mercadé Melé, P. (2022a). How tourists' animosity leads to travel boycott during a tumultuous relationship. *Journal of Tourism Recreation Research*. https://doi.org/10.1080/02508281.2022.2124023

Siyamiyan, A., Hosseini, S., Garcia, F. A., & Macias, R. C. (2022b). Complexities of women solo travelling in a conservative post-Soviet Muslim society: The case of Uzbek women. In C. M. Hall, S. Seyfi, & S. M. Rasoolimanesh (Eds.), *Contemporary Muslim travel cultures* (pp. 155–169). Routledge.

Stern, P. C., Dietz, T., & Kalof, L. (1993). Value orientations, gender, and environmental concern. *Environment and Behavior*, *25*(5), 322–348.

Tallontire, A., Rentsendorj, R., & Blowfield, M. (2001). *Ethical consumers and ethical trade: A review of current literature policy series*. University of Freenwich.

Veluchamy, R., Bharadwaj, M. V., Vignesh, S., & Sharma, G. (2016). Personal and professional attitudes of Generation Z students. *International Journal of Circuit Theory and Applications*, *9*(37), 471–478.

Walters, P. (2021). Are Generation Z ethical consumers? In N. Stylos, R. Rahimi, B. Okumus, & S. Williams (Eds.), *Generation Z marketing and management in tourism and hospitality* (pp. 303–325). Palgrave Macmillan.

Ward, S. (1974). Consumer socialization. *Journal of Consumer Research*, *1*(2), 1–14.

12 Save it to cherish: the rise of wildlife voluntourism with Generation Z

Marie-Louise Bank and Helena Maria Correia Neves Cordeiro Rodrigues

Introduction

In 2015, approximately 1.186 million people travelled the world (UNWTO, 2016), with about 10 million of those travelling as voluntourists (Popham, 2015). According to Purvis and Kennedy (2016), voluntourism (VT) represents a tiny portion of the tourism industry, but it is still a billion-pound industry that must be regulated and developed. This ever-growing market allows tourists to participate in genuine experiences while directly contributing to local communities (Wong et al., 2014). As the VT industry develops, new-emerging micro-niches, such as wildlife and the environment, become trends (Stainton, 2016). To increase accessibility and publicity for this type of market, many organisations have started to participate in planning and offering volunteering touristic experiences. Worldwide Opportunities in Organic Farms (WWOOF) has been making notable use of VT, which has helped diminish the negative economic impacts on tourism following the coronavirus pandemic (Hossenally, 2020).

Burns et al. (2008) present empirical data indicating that people's motivations are the most significant determinants of volunteering activities. They found that the intentions of volunteering have a significant correlation with age groups because people relating to the same cohorts and social groups usually have mutually associated volunteering motivations (Burns et al., 2008). According to age groups, the lowest percentage of volunteers belongs to 'Generation Z' (Gen Z), which denotes young adults born from the mid-1990s through to 2002. This is especially notable as this age group typically demonstrates a great inclination towards volunteering (Cho et al., 2018).

This chapter's main objective is to examine VT motivations with a link to wildlife VT (WVT), namely the VT participants (Gen Z), the non-participants/prospective participants (Gen Z), and VT operators (VTOs). The research approach is to create semi-structured interviews about motivational factors based on the established literature (Harrell & Bradley, 2009). Leximancer software recognises the responses and creates a conceptual map visualisation that includes themes and concepts for further interpretation (Angus et al., 2013). Finally, an analytic evaluation of these outcomes is undertaken to truly comprehend the determinants influencing Gen Z attitudes and perceptions towards WVT.

DOI: 10.4324/9781003289586-16

Literature review

Voluntourism—a new trend in tourism

VT is a niche market within the tourism industry that has been developed since the 1990s (Lee & Zhang, 2020). According to Wearing (2001), the term VT applies to those vacationers who, for varying reasons, consume vacations that may include supporting or mitigating hardship in various nations, restoring a destination's environment, or examining factors that affect the environment, the society, or the government. VT tours are generally fairly short, typically less than a month (Callanan & Thomas, 2005). India was found to be the top-most VT-project-based nation in a study by Novelli (2005), while China, India, the USA, Indonesia, and Brazil are the five leading nations that provide volunteering opportunities. A typical VT profile entails travellers paying to assist in another place. This distinguishes VT from volunteering, where the primary purpose of the trip is to work and serve for more extended periods (Douglas & Greenhill, 2017).

Uriely et al. (2003) stated that VT has increased the growth of tourists' under-standing of ecology and their responsibilities. The authors defined VT as social work wherein tourists perform voluntary activities in their leisure time (Wearing & McGehee, 2013). This interpretation is in line with the work of McGehee and Santos (2005), who claimed that voluntourists use their leisure time and salaries to travel globally to help societies in need. VT plays a vital role in making tourists aware of the projects of local communities (McGehee & Andereck, 2009). Along with volunteering for people or the community, VT has also started to assume a fundamental part in sustaining, protecting, and conserving wildlife (Rattan et al., 2012). This study has focused on VT concerned with wildlife conservation, which is known as wildlife VT (WVT; Cousins et al., 2009).

Various studies have demonstrated several cases of success in the VT industry. A great example is the case of a for-profit, environmental VT project in Costa Rica (Schneller & Coburn, 2018). It seems that environmental VT is popular among the younger generation, which can be explained by their expected responsibility to maintain the environment. Reinforcing this, there has been a shift in educational content over the years, with Millennials and Gen Z prioritising their education on science and the importance of preserving the environment.

Competition is also increasing among VT tour organisers as volunteering ser-vices emerge for travellers in more tour organisations (Smith & Font, 2014). The development of VT has prompted a move from non-profit firms to commercial-based organisations (Wearing & McGehee, 2013). However, across the literature, there remains a pattern of conflict between the benefits and risks of VT. Neverthe-less, the potential benefits of this touristic model should not be overlooked due to fear and uncertainty. Instead, VT should be supported more than ever by all the individuals and organisations involved, with professional, ethical, and human practices.

Voluntourism motivation

An early study led by Katz (1960) expressed that an individual's motivation is essential to understand the process of change and formation regarding their disposition. The attitude and behaviour of a person are shaped depending on social convictions, which demonstrate that motivations are the most fundamental determinants of people (Hsu & Huang, 2012). Regarding VT, favourable attitudes and behaviours are the results of the volunteers being motivated enough to undertake their trips. Fisher and Price (1991) announced that as volunteers create additional motivations due to their social and inborn persuasive viewpoints, they also acquire positive perspectives that further increase their general fulfilment in the VT experience. Lee et al. (2014) analysed the motivations of volunteers who took part in a VT project and demonstrated that the volunteers' motivations could build their ideal perspectives.

The primary and foremost motivation of volunteers is to assist and aid less fortunate populations. Escaping from daily routines was another factor that has been asserted to motivate voluntourists' engagement in VT (Lo & Lee, 2011). This understanding is supported by an investigation conducted by Carter (2008), who found that helping others was the primary motivation for VT, while the desire to do something new was another motivating factor.

Carpenter and Myers (2010) emphasised that volunteers also satisfy their psychological and social objectives when participating in VT. Brown (2005) posits that the four primary reasons for a person engaging in VT are immersion in the community, giving back, finding camaraderie, and seeking educational and bonding opportunities. Likewise, intrinsic motivations also encourage a person's participation in VT, such as encountering something different or new, discovering another country and culture, living in another nation, expanding one's knowledge, and meeting unfamiliar individuals (Benson & Seibert, 2009). A final reason for escaping from daily routines was another factor that has been asserted to motivate voluntourists' engagement in VT (Lo & Lee, 2011). The Volunteer Functions Inventory (VFI) tool, introduced by Clary et al. (1988), can be used to identify the best possible motivation factor for volunteering tourists.

Broad's (2003) study indicated that when presented with an alternative culture, voluntourists had a changed view of the world, were more liberal, calm, pleased with themselves, more composed, and were less self-centred throughout their experience of VT. A study conducted in Australia on the sociology of VT recruitment in higher education found that the practice of recruiting students for VT "is an example of public pedagogy that reinforces a hegemonic discourse of need" (McGloin & Georgeou, 2016, p. 403). Undeniably, no matter the critics, many individuals are likely to support VT as a means of enriching one's CV while benefiting (or not) from an actual self-enhancing experience depending on the partakers' motivations, experiences, and expectations. Appropriately, a person's qualities and encounters are potential push factors for their empowering cooperation in VT (Mannino et al., 2010).

VT has become a commodity in the tourism business as private and large tour organisations have emerged in the field. However, numerous organisations send volunteers who do not have valuable competencies and capabilities, thereby acting as a deterrent and negatively impacting the achievement of the tour's main goal and outcome (Guttentag, 2009). Furthermore, many of the tasks of the VT that the volunteers take part in can be executed by individuals from the domestic or host community, who may be more talented than volunteers (Ver Beek, 2006). VT can also create pressure and envy in the local community since the VT projects may be advantageous for one local area but not others (Sin, 2010). A study in Nepal by Durham (2016) stated that VT creates more harm than good. The research indicated that even if the intentions of a VT tour are good, the rights of the voluntourists remain unprotected, creating room for potential subsequent damage to the group. Similar research by Freidus (2016) in Malawi revealed that volunteer tourists are often left with a superficial understanding of poverty and culture.

To conclude, Volunteers are inspired for different reasons, both selfless and non-benevolent. Understanding the volunteers' motivations and establishing facilities that address their issues can be viewed as significant for VT organisations. Negative impacts must be further considered in the future to ensure that fewer arise and that procedures are put in place to avoid more negative consequences.

Wildlife voluntourism and sustainable development

Behind VT is the attraction of the idyllic destination (Grimm & Needham, 2012), and as a showcase, it can be employed as a method to attract global volunteers for wildlife conservation programmes (Van Tonder et al., 2017). Lorimer (2010) posited that the motivation of WVT volunteers is to experience close contact with wildlife and animals. Volunteers are encouraged to take on challenging tasks such as wilderness hikes and experiencing unusual situations; these are just some of the live adventures that contribute as motivations to the volunteers of WVT tours and projects. According to Brondo (2019), WVT can be considered an illustration of the neoliberal preservation of wild animals and creatures.

Organisations offering WVT find their programmes to be more profitable and productive as the volunteers for WVT spend considerably more money than those of any other VT domestic project. Furthermore, WVT volunteers generally stay and work for a longer period, allowing them to undertake long-term work on WVT tasks (Alexander, 2012; Guttentag, 2009; Stoddart & Rogerson, 2004).

Sustainable VT (SVT) travels are motivated by volunteering while creating sustainable economic and ecological benefits for the participating VTs. Nonetheless, it has been stated by researchers that in some instances, even when VT occurs, the purposes for sustainable development are still not met (Lee & Zhang, 2020). The Council of Europe (1997) characterised SVT as a travel industry movement that secures, preserves, and conserves nature and wildlife, culture, and social assets. Volunteer-based ecotourism thus concerns regular habitats and takes part in the security of the environment and wildlife (Jegou, 2013; Brightsmith et al., 2008).

The United Nations Environment Programme (UNEP) and UNWTO (2005) highlight that SVT is neither a category nor a sector of the travel industry because the standards of SVT are pertinent to all sectors of the travel industry and all destinations. Wildlife tourism (WT) volunteers participate in activities that sustain, protect, and conserve wildlife. WT and VT attractions often comprise traveller destinations and exercises that focus on providing advantages to the local area and the nearby wildlife (Peattie & Moutinho, 2000). WT prompts sustainable economic benefits for some nations while guaranteeing the preservation of many jeopardised species (Shackley, 1996).

WT additionally assumes a significant role in developing nations (Belicia & Islam, 2018). WT boosts the conservation of animals and their habitats by providing financial incentives. Income is made through the privatisation of common assets into products that can be showcased to travellers.

Sustainable wildlife VT (SWVT) can likewise provide regional networks with the freedom to improve through managing volunteers, income-sharing plans, and natural resources side by side in a way that enhances WT. SWVT utilises ecological assets, and biodiversity, nature, and animals are saved, preserved, and conserved (Knafou, 2007). WT has been rapidly expanding over decades due to tourists and VT (Newsome et al., 2005). It is moving beyond the aim of sustainability tourism (Manfredo et al., 2002).

Generation Z (Gen Z)

Gen Z can genuinely be alluded to as natives of the digital era as they have never known a period without the Internet (Gentina, 2016). On a global scale, the world's population is young, with persons aged 24 and under (i.e., Gen Z) accounting for the majority of the population (UN Department of Economic and Social Affairs (DESA, 2020). Another trait of Gen Z is that they are generally more environmentally aware than people from previous generations. Regarding VT, this situation poses the question of whether Gen Z voluntourists would be willing to surrender certain aspects of comfort to support environmentally sustainable practices (Francis & Hoefel, 2018).

As they are thought to be low spenders, VT organisations often neglect Gen Z among worldwide travellers. Nevertheless, Gen Z plays a substantial role as an influencer of change. Previous research has reported that Gen Z has progressively seen travelling as a fundamental component of their life (UNWTO, 2011). The number of VT volunteers was expected to reach 300 million by 2020 (Haddouche & Salomone, 2018). It is hence essential to determine the factors motivating and influencing Gen Z to participate in VT (Body & Tallec, 2015).

Current literature has demonstrated the rising interest and affectability of Gen Z towards VT. Sherraden et al. (2008) propose that this is potentially due to VT providing the opportunity for Gen Z volunteers to universally enjoy a reprieve from school or work, meet and befriend new individuals, learn abilities, upgrade a resume, or find a new line of work internationally. According to a study of Gen

Z's transformative experiences in educational tourism, this generation is capable of recognising the environment and comprehending the role of tourism in it (Wee, 2019).

As posited by Robinson (2018), any successful tourism industry player requires the ability to recognise change and effectively respond to this change. Generational change is one such occurrence, offering both opportunities and challenges for tourism destinations. To conclude, Gen Zers are compatible with VT as they are prone to focus on self-amusing and self-enhancing experiences rather than destinations. Moreover, this generation enjoys supporting sustainable practices, which are quite common in the VT industry. The parties responsible for the planning and sales of VT trips should thus use digital tools and cultural experiences to increase the flow of Gen Z.

Methodology

To achieve our main goal the most fitting research approach is a qualitative method. (Dilanthi et al., 2002). Semi-structured interviews were conducted to give interviewees a degree of independence in articulating their perspectives and to ensure the optimal delivery of their particular interests and knowledge (Horton et al., 2004).

Data collection

Substantial research was performed on websites, VT blogs, and social media, to identify the most suitable participants. Criterion sampling was applied, and it was, therefore, essential that each interviewee fit into one of the three interview groups (Creswell, 2013).

The qualifications for being categorised into one of the three interview groups are as follows:

a) The VT participant (VTP; Gen Z) born between 1997 and 2002, is from a developed country and has participated in a wildlife volunteer programme abroad for four weeks or longer.
b) The VT non-participant (VTNP; Gen Z) born between 1997 and 2002, is from a developed country and has never participated in a wildlife volunteer programme.
c) The VTO is a provider of WVT and is working with ethical and well-established NGOs and projects in emerging countries.

The interviews were conducted in English and recorded (with participants' consent). Present on the consent form was a segment about the interviewees' demographics, namely gender, age, and nationality, and questions about the location and length of stay of the wildlife volunteering project that they had attended. Every interviewee received a numerical code within their interview group, which granted them complete confidentiality of personal identity (Regulation (EU) 2016/679 of the European Parliament and of the Council of April 27, 2016).

Data population

There were 26 participants (Table 12.1). Eight of these received the qualification of (a) VTP; 11 qualified for (b) VTNP; and seven qualified for (c) VTO. The interviewees' demographics were asked on the consent form as closed questions (Korkeakoski, 2012) and reports that an equal number of interviewees identify as Male and Female. Since it was required that all VTPs and VTNPs in the study were part of Gen Z, the main age group is 18–24; the age group 35–44 is the most frequent for VTOs. There are 11 different nationalities given; the majority (n = 10) are German, followed by Swiss and Zimbabwean participants (three from each country), and the volunteering destinations are all in less economically developed countries.

Data analysis tool

The data were analysed with Leximancer (2021) software, which performs automated textual analyses based on text properties and provides an output in a conceptual map (Rodrigues et al., 2020). The software is a text analytic tool that was used

Table 12.1 Participants' demographics

Participant ID	Gender	Age	Nationality	Destination
Voluntourism Participant (VTP) 1	M	18–24	German	South Africa
Voluntourism Participant 2	M	18–24	German	South Africa
Voluntourism Participant 3	F	18–24	Swiss	South Africa
Voluntourism Participant 4	M	18–24	German	South Africa
Voluntourism Participant 5	M	18–24	American	Costa Rica
Voluntourism Participant 6	F	18–24	German	Indonesia
Voluntourism Participant 7	M	18–24	German	Spain
Voluntourism Participant 8	M	18–24	Spanish	Peru
Voluntourism Non-Participant (VTNP) 1	M	18–24	Italian	-
Voluntourism Non-Participant 2	M	18–24	Swiss	-
Voluntourism Non-Participant 3	F	18–24	Swiss	-
Voluntourism Non-Participant 4	F	18–24	German	-
Voluntourism Non-Participant 5	M	18–24	German	.
Voluntourism Non-Participant 6	F	18–24	American	-
Voluntourism Non-Participant 7	F	18–24	German	-
Voluntourism Non-Participant 8	M	18–24	Japanese	-
Voluntourism Non-Participant 9	M	18–24	Chinese	-
Voluntourism Non-Participant 10	F	18–24	Portuguese	-
Voluntourism Non-Participant 11	M	18–24	Portuguese	-
Voluntourism Operator (VTO) 1	F	18–24	Kenyan	Kenya
Voluntourism Operator 2	F	18–24	Namibian	Namibia
Voluntourism Operator 3	F	45–54	German	Costa Rica
Voluntourism Operator 4	F	45–54	Zimbabwean	Zimbabwe
Voluntourism Operator 5	M	35–44	German	South Africa
Voluntourism Operator 6	F	35–44	Zimbabwean	Zambia
Voluntourism Operator 7	F	35–44	Zimbabwean	Zimbabwe

to analyse the textual documents and the information was displayed in a Concept Map, thus representing the main concepts in the text and their relation, and can be useful for analysing interview transcripts (Leximancer, 2021)

Analysis and results

The conceptual map (Figure 12.1) illustrates the stakeholders' perceptions regarding WVT motivation. The Leximancer output includes various rainbow-coloured themes, along with smaller grey nodes symbolising concepts. The six main themes (as determined by the number of hits) are Volunteers (125), Animals (124), Experience (115), Project (53), Life (51), and Local (18). These are thought to be the most critical factors influencing Gen Z's motivation for WVT.

Discussion

This chapter discusses six core themes related to a) VTP's motivations for engaging in a VT programme; the b) VTNP, and c) the VTOs that emerge from the data analysis tags (Figure 12.1) and incorporates extracts from the three groups interviewed and compares them with the existing literature.

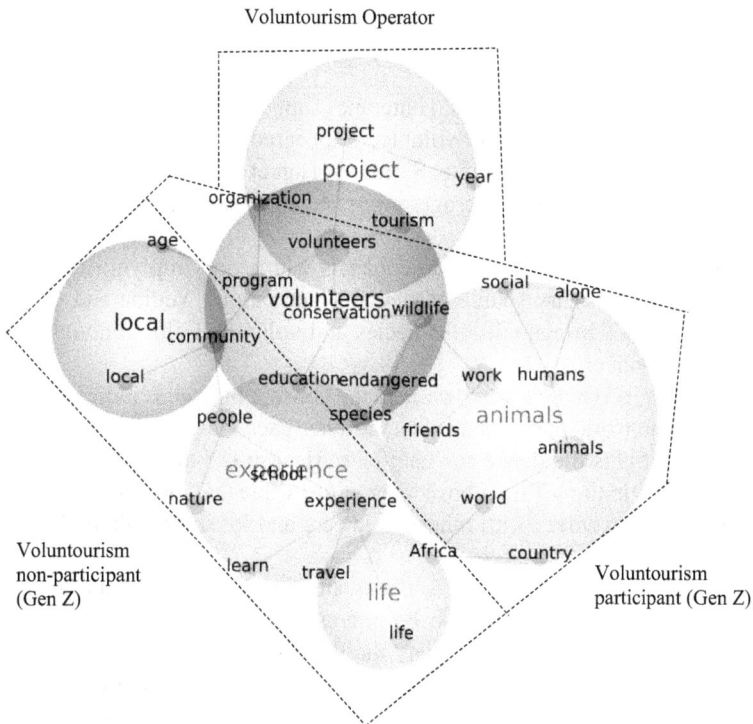

Figure 12.1 Conceptual map illustrating voluntourism motivations with a link to wildlife

The voluntourism participants (VTP)

Wildlife, programme, conservation, and *community* are concepts that connect to the main theme *'Volunteer(ing)'* highlighting the theme's importance for motivational factors of VTPs. Individuals participating in VT pay for the tour in an area where they decide to participate, and this encounter generally includes helping local communities by, for example, improving living conditions, or protecting and ensuring the safety of the environment or endangered species of animals (Polus & Bidder, 2016).

The *wildlife* can emphatically impact sustainability through direct practices, such as maintaining habitats through funds, taking anti-poaching steps, and successfully reintroducing captive-bred animals. Broad and Jenkins (2008) note that VTPs are encouraged by the freedom to study the climate and participate in protection and conservation work. Isiugo and Obioha (2015) have found that endangered species are a significant concern for volunteers and projects implemented by VT associations. VTP #5 supports this by indicating:

> Our average client age is now around 30 years old, and we even have clients older than 60. We do not have a target group based on age or gender but the people being interested in wilderness, nature, and wildlife.
>
> Everyone can learn from each other, and everyone has something to contribute.
>
> (VTP#7)

Taken together, these statements and outcomes support Cho et al.'s (2018) observation that the lowest percentage of volunteers appeared to be Gen Z, which supports the choice of the primary objective. Similarly, Han et al. (2019) reveal that young volunteers who once joined VT programmes become motivated to continue participating in VT programmes.

The second concept of 'A*nimals*' connects with *work*, implying that working with animals is a primary source of Gen Z's motivations. Yudina and Grimwood (2016) say that VT brings wildlife species and volunteers close enough together that they can connect.

Another connection between *work* and the concept of *friends* indicates that volunteers prefer to work "with friends that motivate each other" (VTP #1). Moreover, VTNP #6 highlights the desire to "gain new friendships" through the VT experience. Volunteers in the VT field have a perceived value that incorporates long-term relationship improvement with other volunteers and local individuals (Carvache-Franco et al., 2019).

A link between the concept of *animals* and *humans* leads to the assumption that wildlife volunteers would also be interested in humanitarian work. Participants such as VTP #3 responded that "the bond with animals is more vital than with humans". Even stronger opinions are seen with participants such as VTP #6: "We are destroying our host, our planet. And the animals do not have the voice to speak up", and VTNP #7 adds that it is mankind's "task and job to protect wild animals from interferences in their natural environment, often human-made, such as poaching".

The voluntourism non-participant (VTNP)

The conceptual map illustrates the theme '*Experience*', close to the interviewers of non-participants in VT, which brings together the concepts of *people, travel, and learning*. The concept of *people* connects to the concepts of willingness to help the local *community* and being close to *nature*. VTNP #1 exemplifies this connection by stating that "people who care about nature and the environment want to give back time, resources, and skills to communities".

Non-participants noted that a VT programme provides "an excellent experience after school to learn new skills, to learn from different cultures" (VTNP #5) and a "chance to learn about themselves . . . and most importantly, learn about conservation issues" (VTNP #3). These quotes create a link between *experience* and *learning*. This connection is consistent with Sin's (2009) investigation, which states that prospective volunteers may be inspired to *learn* and develop themselves due to freedom.

The '*Life*' theme includes concepts such as willingness to *travel* and *Africa*. A direct link from *life* connects to the concept and theme *experience*, which implies a close relationship. According to VTNP #3, "*Individual participants are interested in having specific life experiences*".

Moreover, VTNP #6 believes that exposure to "new cultures would influence their lives", and VTNP #5 expects "great benefits for overall life experience". According to Hammersley (2014), volunteers are predisposed to participate in an industry that offers numerous promises for an advancement plan based on support, learning, comprehension, and relationship-building.

The final core theme is '*Local*,' which links to both *community* and the central theme of '*volunteer*'. The non-participant interviews show that they are more worried about the effects of volunteering than those who participated. They want to ensure that local communities are "respected too" and that the communities "see a real benefit from the voluntourism industry" (VTNP #4). VT is therefore an approach to benefit and respect the need to connect with local community projects.

The voluntourism operator (VTO)

We can see that the theme '*Project*' is close to the VTOs and overlaps with the core theme 'volunteer', which highlights the significance of the relationship between the two themes. The *volunteer* concept links *projects* to *tourism*, implying that VT is an approach that benefits volunteers, tourists, and locals (Chen & Chen, 2011). This link also illustrates that VT programmes play a vital role in raising tourists' awareness about local community projects (McGehee & Andereck, 2009). Also, VTO #6 relates the concepts by saying that they want "to create a product that offers people an opportunity to gain conservation awareness, make knowledge and sense of conservation".

According to Wearing (2001), the volunteers' basic characteristics are enthusiasm and autonomy, with no intention of remuneration or personal benefit. This argument is in line with the fact that volunteers bring monetary benefits to VTOs, along with social and climate-related benefits that are not exclusive to the local community (Ver Beek, 2006). Holmes and Smith (2012) add that VT stakeholders see inspiration to meet new individuals, are pleased with their nation of origin, and need to utilise their insights to profit from them, and help trip makers.

Programmes such as those mentioned by VTO #3 are engaged in "working in environmental education programmes for the local youth" to include the desires of Gen Z regarding "supporting the local communities that are affected by the loss of nature" (VTO #2).

Summary and conclusion

Given that the interviews conducted focused on WVT, working with animals is the most significant inspiration for Gen Z volunteers. Moreover, this study has found that VTs feel more motivated when the VT programme includes endangered animals and species. Volunteers are also inspired by close interactions with nature and the conservation goals of their VT projects. According to the interviews, Gen Z believes WVT provides them with more exposure to real-world issues such as climate change and species extinctions caused by mankind. Furthermore, positive interactions with local communities and local issues are another motivator for Gen Z to engage in and continue to participate in VT projects. During a VT experience, Gen Z wants to learn about and live among new cultures, and as a result, they become more aware of their privileges and high standards of living. Furthermore, WVT volunteers are willing to participate in a programme in any country as long as the VTO suits them and their needs.

In contrast, the most significant obstacle to Gen Z engaging in a VT experience is the time required for the programme arrangements. As a result, VTOs need to comprehend these barriers and choose the most successful programmes to attract potential volunteers.

In summary, the main motivations for Gen Z volunteers are the act of volunteering itself; close contact with animals and saving endangered species for future generations; an experience that involves travel; a project that benefits volunteers and locals; a positive effect on the volunteers' lives after the project; and the desire to contribute to local communities.

Contribution to theory

This study advances previous studies that largely concentrate on Generation Y, by highlighting the VT motivations regarding Gen Z specifically. The overlooking of Gen Z in the literature shows that researchers have ignored the significant role of Gen Z in VT projects.

Furthermore, previous literature has for the most part focused on VT in general or only humanitarian volunteering, such as Chen and Chen (2011) and Han et al.

(2019). This research fills a gap in the literature by identifying Gen Z's motivating factors for WVT. Moreover, while interviews with VIPs, VTOs, and local host communities are reported in the literature, interviews have not been performed with VTNPs. This research does so and offers a unique perspective on Gen Z's motivations regarding non-participation (or prospective VT) in volunteering. Ultimately, this research contributes to the VT literature by providing unexplored insights on WVT conservation and sustainability and Gen Z's motivating factors for VT.

Practical contribution

This study offers VTOs and the VT stakeholders (e.g., participants, prospective participants, and operators) an opportunity to adapt strategies in developing or further expanding Gen Z's participation in VT projects.

VTOs and the VT industry should tailor their recruitment materials to appeal to the interests of Gen Z. When volunteers understand the meaning, goals, objectives, tasks, and responsibilities required of them throughout a volunteer project, they feel confident participating. As a result, VTOs and the VT industry should critically revise the participating information to openly communicate what volunteers should expect.

Moreover, VTOs' websites should contain stories and testimonials from former volunteers. VTOs are encouraged to communicate the success of volunteers' efforts. Volunteers must be able to see how their efforts affect the wildlife or conservation project goals.

Offering short programmes with a two-week commitment should be adequate to draw more volunteers. Additionally, participants will likely feel that they have more flexibility if they can choose their arrival and departure times.

Limitations and future research

Although this research has accumulated a substantial amount of information, it has methodological limitations due to the subjective nature of qualitative analysis (Connelly, 2013). This research has not considered factors that might influence a person's motivation to volunteer such as social status (Burns et al., 2008). A future study with participants born after 2002 would be beneficial to track evolving trends and shifting behaviours of future generations, such as Generation Alpha.

Lastly, future research should evaluate the extent of the VT industry's adverse effects on wildlife and the negative aspects of unethical conservation projects. Such studies should include educating volunteer tourists about incorrect approaches, such as breeding wild animals in enclosures to attract tourists and generate more revenue.

Acknowledgement

This work was supported by Fundação para a Ciência e a Tecnologia (FCT), grant UIDB/00315/2020

References

Alexander, Z. (2012). International volunteer tourism experience in South Africa: An investigation into the impact on the tourist. *Journal of Hospitality Marketing and Management*, *21*(7), 779–799.

Angus, D., Rintel, S., & Wiles, J. (2013). Making sense of big text: A visual-first approach for analysing text data using Leximancer and Discursis. *International Journal of Social Research Methodology*, *16*(3), 261–267.

Belicia, T., & Islam, M. (2018). Towards a decommodified wildlife tourism: Why market environmentalism is not enough for conservation. *Societies*, *8*(3), 59.

Benson, A., & Seibert, N. (2009). Volunteer tourism: Motivations of German participants in South Africa. *Annals of Leisure Research*, *12*(3–4), 295–314.

Body, L., & Tallec, C. (2015). *L'experience client*. Eyrolles.

Brightsmith, D. J., Stronza, A., & Holle, K. (2008). Ecotourism, conservation biology, and volunteer tourism: A mutually beneficial triumvirate. *Biological Conservation*, *141*(11), 2832–2842.

Broad, S. (2003). Living the Thai life – a case study of volunteer tourism at the Gibbon Rehabilitation Project, Thailand. *Tourism Recreation Research*, *28*(3), 63–72.

Broad, S., & Jenkins, J. (2008). Gibbons in their midst? Conservation volunteers' motivations at the Gibbon rehabilitation project, Phuket, Thailand. In D. K. Lyons & S. Wearing (Eds.), *Journeys of discovery in volunteer tourism: International case study perspectives* (pp. 72–85). CABI Publishing.

Brondo, K. V. (2019). Entanglements in multispecies voluntourism: Conservation and Utila's affect economy. *Journal of Sustainable Tourism*, *27*(4), 590–607.

Brown, S. (2005). Travelling with a purpose: Understanding the motives and benefits of volunteer vacationers. *Current Issues in Tourism*, *8*(6), 479–496.

Burns, D. J., Reid, J., Toncar, M., Anderson, C., & Wells, C. (2008). The effect of gender on the motivation of members of Generation Y college students to volunteer. *Journal of Nonprofit and Public Sector Marketing*, *19*(1), 99–118.

Callanan, M., & Thomas, S. (2005). Volunteer tourism deconstructing volunteer activities within a dynamic environment. In M. Novelli (Ed.), *Niche tourism: Contemporary issues, trends and cases* (pp. 183–200). Butterworth-Heinemann.

Carpenter, J., & Myers, C. K. (2010). Why volunteer? Evidence on the role of altruism, image, and incentives. *Journal of Public Economics*, *94*(11–12), 911–920.

Carter, K. A. (2008). *Volunteer tourism: An exploration of the perceptions and experiences of volunteer tourists and the role of authenticity in those experiences* [Masther's thesis, Lincoln University]. https://researcharchive.lincoln.ac.nz/bitstream/handle/10182/526/carter_mapplsc.pdf?sequence=7&isAllowed=y

Carvache-Franco, M., Carvache-Franco, W., Contreras-Moscol, D., Andrade-Alcivar, L., & Carvache-Franco, O. (2019). Motivations and satisfaction of volunteer tourism for the development of a destination. *Geo Journal of Tourism and Geosites*, *26*(3), 714–725.

Chen, L. J., & Chen, J. S. (2011). The motivations and expectations of international volunteer tourists: A case study of "Chinese village traditions". *Tourism Management*, *32*(2), 435–442.

Cho, M., Bonn, M. A., & Han, S. J. (2018). Generation Z's sustainable volunteering: Motivations, attitudes and job performance. *Sustainability*, *10*(5), 1400.

Clary, E. G., Snyder, M., Ridge, R. D., Copeland, J., Stukas, A. A., Haugen, J., & Miene, P. (1988). Understanding and assessing the motivations of volunteers: A functional approach. *Journal of Personality and Social Psychology*, *74*(6), 1516–1530.

Connelly, L. M. (2013). Limitation section. *MEDSURG Nursing, 22*(5), 325–336.

Council of Europe. (1997). *Naturopa, 84–1997.* https://rm.coe.int/naturopa-1997-no-84/
168069cdf2

Cousins, J. A., Evans, J., & Sadler, J. P. (2009). "I've paid to observe lions, not map roads!" –
An emotional journey with conservation volunteers in South Africa. *Geoforum, 40*(6),
1069–1080.

Creswell, J. (2013). *Research design: Qualitative, quantitative, and mixed methods
approaches.* Sage.

Dilanthi, A., David, B., Marjan, S., & Rita, N. (2002). Quantitative and qualitative research
in the built environment: Application of "mixed" research approach. *Work Study, 51*(1),
17–31.

Douglas, T., & Greenhill, A. (2017). What is voluntourism? *Interaction, 45*(1), 33–35.

Durham, J. (2016). *Protecting the voluntoured: An explanatory human rights impact assessment for ethical voluntourism in Nepal* [Master's thesis, The University of Sydney].
https://repo.gchumanrights.org/bitstream/handle/20.500.11825/254/Durham_GC_Asia-
Pacific%28APMA%29_15-16.pdf?sequence=1&isAllowed=y

Fisher, R. J., & Price, L. L. (1991). International pleasure travel motivations and post-vaca-
tion cultural attitude change. *Journal of Leisure Research, 23*(3), 193–208.

Francis, T., & Hoefel, F. (2018). *"True Gen": Generation Z characteristics and its implications for companies.* McKinsey and Company. www.mckinsey.com/industries/consumer-
packaged-goods/our-insights/true-gen-generation-z-and-its-implications-for-companies

Freidus, A. L. (2016). Unanticipated outcomes of voluntourism among Malawi's orphans.
Journal of Sustainable Tourism, 25(9), 1306–1321.

Gentina, É. (2016). *Marketing et Génération Z: Nouveaux modes de consommation et stra-
tégies de marque.* Dunod.

Grimm, K. E., & Needham, M. D. (2012). Moving beyond the "I" in motivation: Attributes and perceptions of conservation volunteer tourists. *Journal of Travel Research, 51*(4),
488–501.

Guttentag, D. A. (2009). The possible negative impacts of volunteer tourism. *International
Journal of Tourism Research, 11*(6), 537–551.

Haddouche, H., & Salomone, C. (2018). Generation Z and the tourist experience: Tourist
stories and use of social networks. *Journal of Tourism Futures, 4*(1), 69–79.

Hammersley, L. A. (2014). Volunteer tourism: Building effective relationships of understanding. *Journal of Sustainable Tourism, 22*(6), 855–873.

Han, H., Meng, B., Chua, B.-L., Ryu, H. B., & Kim, W. (2019). International volunteer
tourism and youth travelers – an emerging tourism trend. *Journal of Travel & Tourism
Marketing, 36*(5), 549–562.

Harrell, M. C., & Bradley, M. A. (2009). *Data collection methods: Semi-structured interviews and focus groups.* www.rand.org/pubs/technical_reports/TR718.html

Holmes, K., & Smith, K. (Eds.). (2012). *Managing volunteers in tourism: Attractions, destinations and events.* Routledge.

Horton, J., Macve, R., & Struyven, G. (2004). Qualitative research: Experiences in using
semi-structured interviews. In C. Humphrey & B. Lee (Eds.), *The real life guide to
accounting research* (pp. 339–357). Elsevier.

Hossenally, R. (2020). Local voluntourism in MidCoast Maine: Redefining travel in Covid
times. *Forbes.* www.forbes.com/sites/rooksanahossenally/2020/06/23/local-voluntour
ism-in-midcoast-maine-redefining-travel-in-covid-times/?sh=2863d5e369dd

Hsu, C. H. C., & Huang, S. (2012). An extension of the theory of planned behavior model
for tourists. *Journal of Hospitality and Tourism Research, 36*(3), 390–417.

Isiugo, P. N., & Obioha, E. E. (2015). Community participation in wildlife conservation and protection in Oban Hills area of Cross River State, Nigeria. *Journal of Sociology and Social Anthropology*, *6*(2), 279–291.

Jegou, A. (2013). Vers un tourisme durable dans lamétropole parisienne? *Métropolisation et tourisme*, 86–98.

Katz, D. (1960). The functional approach to the study of attitudes. *Public Opinion Quarterly*, *24*(2), 163–204.

Knafou, R. (2007). Tourisme et développement durable. In Y. Veyret (Ed.), *Le développement durable SEDES* (pp. 178–192). Armand Colin.

Korkeakoski, L. (2012). *Does voluntourism fulfill the criteria of sustainable tourism?* Kajaani University of Applied Sciences.

Lee, C. K., Reisinger, Y., Kim, M. J., & Yoon, S. M. (2014). The influence of volunteer motivation on satisfaction, attitudes, and support for a mega-event. *International Journal of Hospitality Management*, *40*, 37–48.

Lee, H. Y., & Zhang, J. J. (2020). Rethinking sustainability in volunteer tourism. *Current Issues in Tourism*, *23*(14), 1820–1832.

Leximancer. (2021). *Leximancer user guide: Vol. release 4* (pp. 1–136). Leximancer.

Lo, A. S., & Lee, C. Y. S. (2011). Motivations and perceived value of volunteer tourists from Hong Kong. *Tourism Management*, *32*(2), 326–334.

Lorimer, J. (2010). International conservation "volunteering" and the geographies of global environmental citizenship. *Political Geography*, *29*(6), 311–322.

Manfredo, M. J., Pierce, C. L., & Teel, T. L. (2002). Participation in wildlife viewing in North America. In M. J. Manfredo (Ed.), *Wildlife viewing in North America: A management planning handbook* (pp. 25–32). Oregon State University Press.

Mannino, C. A., Snyder, M., & Omoto, A. M. (2010). Why do people get involved? Motivations for volunteerism and other forms of social action. In D. Dunning (Ed.), *Social motivation* (pp. 127–146). Psychology Press.

McGehee, N. G., & Andereck, K. (2009). Volunteer tourism and the "voluntoured": The case of Tijuana, Mexico. *Journal of Sustainable Tourism*, *17*(1), 39–51.

McGehee, N. G., & Santos, C. A. (2005). Social change, discourse and volunteer tourism. *Annals of Tourism Research*, *32*(3), 760–779.

McGloin, C., & Georgeou, N. (2016). "Looks good on your CV": The sociology of voluntourism recruitment in higher education. *Journal of Sociology*, *52*(2), 403–417.

Newsome, D., Dowling, R. K., & Moore, S. A. (Eds.). (2005). *Wildlife tourism*. Channel View Publications.

Novelli, M. (2005). *Niche tourism: Contemporary issues, trends and cases*. Butterworth-Heinemann.

Peattie, K., & Moutinho, L. (2000). The marketing environment for travel and tourism. In L. Moutinho (Ed.), *Strategic management in tourism* (pp. 17–37). CABI Publishing.

Polus, R. C., & Bidder, C. (2016). Volunteer tourists' motivation and satisfaction: A case of Batu Puteh Village Kinabatangan Borneo. *Procedia – Social and Behavioral Sciences*, *224*, 308–316.

Popham, G. (2015). Boom in "voluntourism" sparks concerns over whether the industry is doing good. *Reuters*. www.reuters.com/article/us-travel-volunteers-charities-idUSKCN0 P91AX20150629

Purvis, K., & Kennedy, L. (2016). Volunteer travel: Experts raise concerns over unregulated industry. *The Guardian*. www.theguardian.com/global-development-professionals-network/2016/jan/13/concerns-unregulated-volunteer-tourism-industry

Rattan, J. K., Eagles, P. F. J., & Mair, H. L. (2012). Volunteer tourism: Its role in creating conservation awareness. *Journal of Ecotourism, 11*(1), 1–15.

Regulation (EU) 2016/679 of the European Parliament and of the Council of April 27, 2016 on the protection of natural persons with regard to the processing of personal data and on the free movement of such data, Official Journal of the European Union 16. (2016). https://eur-lex.europa.eu/legal-content/EN/TXT/PDF/?uri=CELEX:32016R0679&from=ES

Robinson, V. M. (2018). *A new face of tourism : Understanding travel experiences of New Zealand inbound Generation Z* [Master's thesis, AUT University]. http://orapp.aut.ac.nz/handle/10292/12140 https://openrepository.aut.ac.nz/bitstream/handle/10292/12140/RobinsonVM.pdf?sequence=3&isAllowed=y

Rodrigues, H., Brochado, A., & Troilo, M. (2020). Listening to the murmur of water: Essential satisfaction and dissatisfaction attributes of thermal and mineral spas. *Journal of Travel & Tourism Marketing, 37*(5), 649–661.

Schneller, A. J., & Coburn, S. (2018). For-profit environmental voluntourism in Costa Rica: Teen volunteer, host community, and environmental outcomes. *Journal of Sustainable Tourism, 26*(5), 832–851.

Shackley, M. L. (1996). *Wildlife tourism*. International Thomson Business Press.

Sherraden, M. S., Lough, B., & McBride, A. M. (2008). Effects of international volunteering and service: Individual and institutional predictors. *Voluntas: International Journal of Voluntary and Nonprofit Organizations, 19*(4), 395–421.

Sin, H. L. (2009). Volunteer tourism – "involve me and I will learn"? *Annals of Tourism Research, 36*(3), 480–501.

Sin, H. L. (2010). Who are we responsible to? Locals' tales of volunteer tourism. *Geoforum, 41*(6), 983–992.

Smith, V. L., & Font, X. (2014). Volunteer tourism, greenwashing and understanding responsible marketing using market signalling theory. *Journal of Sustainable Tourism, 22*(6), 942–963.

Stainton, H. (2016). A segmented volunteer tourism industry. *Annals of Tourism Research, 61*, 256–258.

Stoddart, H., & Rogerson, C. M. (2004). Volunteer tourism: The case of habitat for humanity South Africa. *GeoJournal, 60*(3), 311–318.

UN Department of Economic and Social Affairs (DESA). (2020). *World: Population in 2020*. https://population.un.org/wpp/

UNEP & UNWTO. (2005). *Making tourism more sustainable – a guide for policy makers*. www.e-unwto.org/doi/book/10.18111/9789284408214

UNWTO. (2011). The power of youth travel. *The World Youth Student and Educational Travel Confederation*, 1–38. www.unwto.org/archive/middle-east/publication/power-youth-travel

UNWTO. (2016). *UNWTO tourism highlights*. www.unwto.org

Uriely, N., Reichel, A., & Ron, A. (2003). Volunteering in tourism: Additional thinking. *Tourism Recreation Research, 28*(3), 57–62.

Van Tonder, S., Hoogendoorn, G., & Block, E. (2017). Conservation volunteer tourism in the Hartbeespoort Region, South Africa: An exploratory study. *African Journal of Hospitality, Tourism and Leisure, 6*(1), 1–13.

Ver Beek, K. A. (2006). The impact of short-term missions: A case study of house construction in Honduras after Hurricane Mitch. *Missiology, 34*(4), 477–495.

Wearing, S. (2001). *Volunteer tourism: Experiences that make a difference*. CABI Publishing.

Wearing, S., & McGehee, N. G. (2013). Volunteer tourism: A review. *Tourism Management*, *38*, 120–130.

Wee, D. (2019). Generation Z talking: Transformative experience in educational travel. *Journal of Tourism Futures*, *5*(2), 157–167.

Wong, J., Newton, J. D., & Newton, F. J. (2014). Effects of power and individual-level cultural orientation on preferences for volunteer tourism. *Tourism Management*, *42*, 132–140.

Yudina, O., & Grimwood, B. S. R. (2016). Situating the wildlife spectacle: Ecofeminism, representation, and polar bear tourism. *Journal of Sustainable Tourism*, *24*(5), 715–734.

Part V

Conclusions

Gen Z and the future of tourism

13 Conclusions and futures

Are Gen Z the sustainable consumers of the future?

Siamak Seyfi, C. Michael Hall, and Marianna Strzelecka

Introduction

This chapter provides an outline of some of the emerging issues that arise in examining the relationship between Gen Z, tourism, and sustainability. Gen Z is poised to become, if it is not already, the world's largest travelling generation. Given that the consumption characteristics of a generation, by definition, have the potential to last a lifetime, understanding the Gen Z market is not only important from the perspective of industry, but has implications for the entire planet. Effective responses to environmental crises such as biodiversity loss, global heating, deforestation, and desertification cannot be developed unless we understand both the contributions of tourism to environmental change and the tourism practices of Gen Z. As the various chapters in this volume attest to, although Gen Z is often portrayed as being the most environmentally friendly generation, the reality is far more complex, with there often being significant gaps between attitudes and behaviours and with the nature of pro-environmental behaviour often varying depending on the type of consumption and the context within which consumption occurs.

Nevertheless, there is a range of research themes with respect to Gen Z, tourism and sustainability that have yet to be built upon and which would help develop a more complete view of behaviour. Some of the issues, as with many generational studies, include the importance of cross-cultural studies as well as paying attention to different economic circumstances. In addition, more attention may need to be given to the role of gender in Gen Z consumption practices. And although long-term generational studies are important, the Sustainable Development Goals (SDGs) provide an immediate focus for research until 2030 given their significance for society, business, and the environment (Hall, 2019, 2022; Rasoolimanesh et al., 2020).

The Muslim Gen Z travel market

There is a great need to extend Gen Z studies out to different countries and cultures. It is estimated that of the world's two billion Gen Z population about one in four (553 million) are Muslim (Mastercard Crescent-Rating, 2023). With an average age of 23.7 years younger than the global average, Muslims are projected to have

DOI: 10.4324/9781003289586-18

the fastest growth rate of all major religious groups (Lipka, 2016). Millennial and Gen Z travellers are expected to reach peak earning, spending, and travelling life phases, making Muslim tourism a highly attractive segment (Vargas-Sanchéz & Perano, 2018; Oktadiana et al., 2020). Muslims under the age of 30 make up approximately 60% of the population in Muslim-majority countries. When it comes to Muslim travel, the younger Muslim population is viewed as a lucrative market (Vargas-Sanchéz & Perano, 2018; Hall & Prayag, 2019; Hall et al., 2023a). The report of Crescent-Rating, the world's leading authority on halal-friendly travel, estimates that more than 30% of Muslim travellers in 2016 were Millennials, while another 30% were Gen Z. In 2016, of the 121 million Muslim international tourist arrivals, over 72 million were Millennials or Gen Z (Tayao-Juego, 2017). Master-card Crescent-Rating (2023) estimates that of the 160 million Muslim travellers in 2019, 37 million were Gen Z. Two key trends are identified as driving Muslim-friendly travel overall: the Millennial generation and Gen Z, and technology that will increase access to travel information, especially with respect to halal tourism and hospitality offerings:

> Destinations and service providers need to ensure that their brands adapt to become well connected to the environments of Gen Y and Gen Z. As the internet and social environments play key roles in their everyday lives, the Muslim travel industry must evolve their offerings to ensure that their brands are reintroduced to these new segments and that their Muslim-friendly services are Authentic, Affordable and Accessible to these young segments.
>
> (CrescentRating, 2017, p. 7)

> Muslim Gen Zs represent 27.2% of the global Muslim population. They are actively redefining what it means to travel. They will lead the narrative of the next phase of the development of Halal travel and will play a crucial role in championing sustainability initiatives.
>
> (Mastercard Crescent-Rating, 2023, p. 4)

This highlights that halal tourism businesses and destinations as well as the wider tourism industry must therefore understand the expectations and behaviours of this growing cohort of travellers. Vargas-Sanchéz and Perano (2018) examined halal tourism from the perspective of Gen Z in Indonesia, a Muslim-majority country, and concluded that more attention is required from scholars to gain new insights into it and to understand aspects of Gen Z consumption from different cultural bases. In their study on Gen Z in Bangladesh, Polas et al. (2022) examined which factors motivate Gen Z to revisit halal restaurants and demonstrated that service quality, physical environment, and price perception positively and significantly influence Gen Z's revisit intention. Such information is important for Islamic branding and the intention of Gen Z to revisit and for halal businesses' knowledge of Gen Z customer preferences and retain customer loyalty, thereby maintaining their advantage in an increasingly competitive business environment. However, while there is a growing literature on Muslim Gen Z consumer behaviour in tourism and hospitality there is more limited attention given to sustainability concerns.

In a study conducted by Mastercard Crescent-Rating (2023) on Muslim Gen Z, it is reported that Muslim Gen Z travellers are prioritising cultural immersion and sustainability in their travels, with 45% of respondents in the survey identifying themselves as environmental enthusiasts and social activists (Dudekula, 2023). Around 11% also said they will pay for carbon offsets and 14% will take holidays closer to home to avoid long-haul flights, while social activism focuses on community tourism initiatives and fair wages in tourism and hospitality businesses. When it comes to travel spend, Muslim Gen Z appear to be intentional spenders with 77% of respondents being willing to spend more on sustainable practices such as reducing air travel, participating in volunteer tourism and supporting local businesses. When it comes to payment methods overseas, 73% of them are more inclined to use a debit or credit card while 57% prefer to use cash. Over 60% of respondents indicate that their travel inspiration comes from social media channels (especially TikTok and Instagram), the same degree of influence as family and friends. More than half (55%) rely on travel blogs, influencers, and 42% look at travel review sites (Mastercard Crescent-Rating, 2023).

Muslim Gen Z women also emerged in the study as important decision-makers, with over 70% heavily involved in planning for family travel. Almost 70% prefer cultural immersion activities such as experiencing local traditions, heritage, and cuisine. In addition, 63% seek opportunities to learn something new in their travels. This segment is also a driver for change, with 76% indicating social causes to be important in their travel plans. However, they also share the wider Gen Z attribute of connectedness as more than 80% of Muslim female respondents stated that they need social media during their travels (Mastercard Crescent-Rating, 2023). An earlier report released in 2019 by Mastercard and Crescent-Rating, stated that Muslim women travellers made up 63 million of 140 million total Muslim visitor arrivals in 2018 with over half of them using some form of social media to scope out accommodation, logistics and dining. Some 28% of Muslim women's journeys in 2018 were solo travel, indicative of a growing younger demographic willing to experience the world and one which is likely to expand into the future (Nikjoo et al., 2021; Emekli et al., 2023; Hall et al., 2023b; Kervankiran et al., 2023).

Gen Z and solo female travel

The solo travel market is believed to be increasingly driven by the Gen Z population (Sebova et al., 2021; Otegui-Carles et al., 2022). For instance, a study conducted by Hostelworld found that Gen Z females are increasingly choosing to travel alone, and bookings made by female solo travellers have increased by 88% since 2015 (Southan, 2019). Significantly, as noted earlier, female solo travel is also becoming more important for Gen Z Muslim women as Islamic consumption becomes more globalised (Hall et al., 2023b). In another survey from Booking.com, independence was a top priority for Gen Z and a third of Gen Z travellers globally (33%) prefer to travel alone (more than any other demographic). Liu et al. (2022) examined the effects of social media marketing activities on Gen Z travel behaviours and compared Gen Z tourists with other generations and analysed gender differences within the Gen Z cohort. The authors reported that Gen Z females were more likely to be influenced by customisation and word-of-mouth features, while Gen Z males were more sensitive to the entertainment features of social media marketing activities. However, while the growth of solo female travel in the future appears clear, its implications for sustainability behaviours are, as yet, unresolved. There are some suggestions that female travellers may be more environmentally and socially active (Nikjoo et al., 2021) and more interested in transformative and low-impact tourism (Pung et al., 2020), but more research is needed given that this may have significant implications for the development of sustainable behaviour interventions (Sebova et al., 2021; Otegui-Carles et al., 2022).

Gen Z and Sustainable Development Goals

Gen Z are often closely associated with action to promote sustainability because of the actions of people like Greta Thunberg as well as the various global climate strike actions involving hundreds of thousands of young people around the world to protest government inaction on climate change. For some commentators, such activities suggest that Gen Z are passionate about the environment and are willing to take bold action to make a difference especially given states of eco-anxiety (Nugent, 2019), while others may raise questions as to how representative such measures are (Plautz, 2020). Nevertheless, these actions are often framed as being typical of the stance of Gen Z with respect to sustainability (Singh et al., 2022; D'Arco et al., 2023). As Gen Z becomes more aware of environmental, social, and economic issues, companies and destinations are being called upon to transform their business models and integrate sustainability principles into their everyday operations (Singh et al., 2022). The study of Gomes et al. (2023) on Portuguese Gen Z demonstrated that environmental concerns, green future estimation and green perceived quality are potential determinants of Gen Z's consumption of green products and positively influence willingness to pay more for green products.

Younger generations are often described as more socially conscious and more likely to promote SDGs than older generations (Yamane & Kaneko, 2021) and are widely regarded as more progressive when it comes to sustainability and sustainability-related lifestyles. However, academic studies that examined generational differences found contradictory results, and most of them do not reveal significant generational differences (Prayag et al., 2022). In contrast, Yamane and Kaneko (2021) examined whether the younger generation is a driving force for endorsing the SDGs by examining generational differences in lifestyle and the job preferences of younger people in Japan. They found that Japanese Gen Z support the SDGs more than older generations and they are willing to sacrifice money to work for enterprises that are committed to SDGs. They also predicted that the younger generation will make up the majority of the labour force by 2030 and will likely play an important role in achieving the SDGs. In another study, on the SDGs for

Gen Z, Karpovic and Eminens (2018) suggested that education on the SDGs should involve at least three actions to be more appealing to Gen Z's needs and lifestyle:

1. integrating SDGs into familiar and interesting themes: popular cartoons and online games;
2. delivering on-demand access: responsive and interactive digital platform with constant gamification; and
3. engaging individuals with challenges: creating various exercises around the 17 SDGs.

However, some SDGs clearly have the capacity to be more attractive to members of Gen Z than others. Based on a survey of more than 1,200 Gen Z's views on SDGs in Australia, China, Hong Kong and Singapore, finding decent work and economic growth (Goal 8), reducing inequality (Goal 10) and climate action (Goal 13) were Gen Z's top sustainability concerns (Sandpiper Communications, 2020). In terms of the pillars of the SDGs, Gen Zs across all these countries were most concerned by environmental issues (86%), followed by social issues (84%), and economic issues (83%). The report also indicated that although economic stability is a concern in the aftermath of the COVID-19 pandemic, environmental issues continue to be of more concern to Gen Z than social and economic issues with climate change as a key environmental concern for Gen Z across these countries (Sandpiper Communications, 2020). Nevertheless, when it comes to actual travel decisions and tourism practices, what may be interpreted as environmentally appropriate and sustainable behaviour may prove to be debatable or even problematic (Prayag et al., 2022). For example, of particular significance to this study, a 2020 study of university students (106 of whom were between the ages of 18 and 25 who are within the Gen Z cohort) on tourism consumption, found that the majority of respondents claimed that travel is crucial to their happiness and a common response was that only cost or legislation would discourage them from flying (Roberts, 2020, cited in Sharpley, 2021).

Gen Z, lifestyle, and political and ethical consumerism

Being born into a generation that has grown up with the Internet and digital media, together with the dramatic sociocultural and political changes that have occurred during their lives, has influenced the development of a Gen Z cohort that, in terms of attitude, appears passionate about a wide range of global issues such as climate change, social justice, and sustainability (Dabija et al., 2019; Djafarova & Foots, 2022). According to Nonomura (2017), citizens, especially young people, are becoming increasingly aware of 'the politics behind products' and the 'complex social and normative context' (i.e., late capitalism, neoliberalism, economic globalisation) in which production and day-to-day consumption occurs" (p. 236). Research suggests that younger generations are becoming more interested in seeking out and paying more for ethical products and services and cognitively interpret their political, environmental, and social concerns into expressed purchase behaviour (Chatzopoulou & de Kiewiet, 2021; Djafarova & Foots, 2022, Seyfi et al., 2023; Gomes et al., 2023). In their study on the drivers underpinning buycott behaviours of Gen Z, Seyfi et al. (2023) explored why Gen Z engages in tourism-related buycotts. The research demonstrated that fulfilment, self-identity, and frugality are core individual drivers of buycotting for Gen Z tourism consumers. The study also found that Gen Z's buycott behaviour is encouraged by prosocial drivers including altruism, trust, and the pursuit of social justice motivations. Furthermore, exposure to social media information, peer persuasion, and prior experience were identified as the influencers of buycott participation. In another study focused on Gen Z, Seyfi et al. (2021) investigated how digital media engagement drives the construction and mobilisation of political consumerism among the Gen Z cohort of travellers. Their study illustrated how digital media enables sustainability-driven political consumerism. Digital media and social media networks were identified as key reinforcers of engagement in tourism-related boycotting and buycotting behaviours as expressions of political consumerism among Gen Zers.

A growing number of studies indicate that younger people are more likely to adopt a lifestyle based on the principles of sustainability (e.g., Dabija et al., 2020; LoDuca, 2020; Djafarova & Foots, 2022; Seyfi et al., 2023; Gomes et al., 2023). For instance, in their study focusing on Gen Z and the retailing sector in Romania, Dabija et al. (2020) reported that Gen Z have a strong interest in sustainable development and social responsibility and are more likely to participate in environmental protection activities. Similarly, Last (2014) showed that Gen Z tend to purchase products from companies that apply sustainability principles and want to reduce their carbon footprint. However, this does not necessarily mean that Gen Z is necessarily willing to adopt new sustainable products. Instead, Gen Z appear to have a strong preference for firms that are deemed ethical and that are aligned with their values. LoDuca (2020), for example, claim that the Gen Z cohort wants firms and businesses to use their power to push for responses to environmental change, civil justice, human rights, and diversity and inclusion.

Gen Zers appear to have a strong preference for firms that are deemed ethical and that are aligned with their values. With their substantial purchasing power, Gen Z is the fastest-growing market group, with distinct purchasing values and the financial ability to act on them (Seyfi et al., 2022, 2023). Studies suggest that when it comes to shopping and purchasing decisions, Gen Z strive to learn about the brand they are purchasing. For instance, the survey conducted by Green Match (2020) indicates that Gen Z is more prepared than previous generations to refuse to buy from businesses that do not suit their standards, even boycotting them. Their survey showed that 40% of Gen Z respondents have boycotted a company in the past, with another 49% considering it in the future. The Masdar Gen Z Global Sustainability Survey (2016) also has some interesting results with respect to Gen Z as it also suggests that they view consumption through an ethical lens and are prepared to boycott companies which are not sustainable. The findings also show that Gen Zers are not only aware of the world's environmental problems, but also believe that poverty, inequality, unemployment, and the economy are similarly critical issues that will become profoundly significant over their lifetimes and for which they are looking to business and governments to respond.

The future of Generation Z's approach to environmental problems and how it affects their sustainable consumption practices in relation to tourism will become a significant area of research in the coming years. Gen Zers will be the first generation who will bear the full brunt of the implications of climate change and global environmental change, including biodiversity loss. A key issue will be the extent to which their contemporary online activism will be sustained over time and transferred to more active boycotts and buycotts. Such measures will clearly have implications not only for companies and destinations but also for travel practices. But what is unclear is the extent to which the positive attitudes towards sustainability in tourism become transformed into particular types of tourism practices, especially with respect to paying for the real costs of travel, and whether they will be sustained over the long term. Given this context there are therefore a number of research challenges faced in researching Gen Z in the future:

- Longitudinal studies are needed for comparing sustainable consumption among different generations and the associated behavioural change over time (Prayag et al., 2022) as well as the barriers to tourism-related ethical and political consumerism. Comparison of pro-environmental and responsible consumption practices of Gen Z tourists would also be valuable between developed countries and developing countries.
- Research on ethical and responsible consumption of Gen Z has mainly focused on economically developed countries (across North America, Europe, and Asia). However, ethical consumption varies significantly among societies and in different cultures and consumers have different motives and values driving ethical consumption (Sudbury Riley et al., 2012). Therefore, cross-cultural studies are essential along with generational studies in less developed countries.
- The link between digital media engagement and the ethical consumption of Gen Z is significant in terms of how digital media use influences ethical consumption (Seyfi et al., 2022) and, related to this, how is this connected to the wider drivers of Gen Z ethical tourism consumerism and practices.
- Another concern with Gen Z, which is potentially related to social media and online practices is that of mental health. Therefore, what connections exist between the mental and behavioural health of Gen Z and their travel behaviour, especially with respect to sustainable tourism?
- From a supply-side perspective, further knowledge is required on how to do businesses, destinations, and the government's marketing strategies embrace Gen Z travellers and what are the implications for sustainable tourism behaviour. There is a need to also understand how businesses, destinations and governments respond to digital political consumerism in terms of their own online tactics and means to engage and counter boycotts and buycotts. This means studying the drivers of political consumerism in tourism and the role of digital media in shaping these drivers, given that the influence of digital media is only

likely to increase in the future, including the role of new virtual forms of digital media.

• Finally, there is a need for long-term analysis of the underlying social and cultural values of Gen Z as political consumers while also seeking to identify the long-term relationships between political consumption and activism in tourism with other forms of political consumption as well as wider sets of political values and actions.

Conclusion

Gen Z present something of a quandary when it comes to understanding if they will become a sustainable generation, especially with respect to tourism. On the surface, much academic and business research suggests that they hold substantial pro-environmental values and attitudes. However, while Gen Z values and attitudes are undoubtedly significant, they do not automatically translate into behaviours that are empirically sustainable. Most studies on Gen Z and sustainable tourism focus on attitudes or intended behaviours rather than actual behaviour. This is problematic because notions of what constitutes green or sustainable behaviours are often weakly constituted in the tourism and hospitality literature and, perhaps even more seriously, the actual contribution to the environment or sustainability is usually never measured. Furthermore, although research on Gen Z and sustainability is usually presented in a positive light with respect to the SDGs actual studies that look at the specific support of Gen Z for the different SDGs are extremely limited. Again, general pro-environmental and sustainability attitudes and values are interpreted as supporting specific behaviours without the necessary supporting evidence. So far there is little empirical evidence to show that there is a positive change. An overview of the Gen Z and sustainability literature would suggest that they want change and see it as highly desirable, but such changes are very much 'within the system' where social entrepreneurship and political consumerism are perceived as 'radical' responses (Hall, 2022; Seyfi et al., 2023).

In conclusion, we are therefore left with a situation in which the tension between growth, consumption and sustainability remains expressed in Gen Z discourses associated with so-called weak sustainability in which sustainability is regarded as being achievable within existing paradigms of growth and consumption, often with a focus on better or more efficient consumption, especially via the use of new information technologies and digital connectedness (Seyfi et al., 2022). In contrast, strong sustainability, that is, more fundamental changes in consumption practices, that Gen Z and following generations likely urgently require for their own futures continues to reside in the background of existing Gen Z and sustainability research. It may therefore still be for future generations to address the central problems of sustainability rather than the consumptive practices of Gen Z.

References

Chatzopoulou, E., & de Kiewiet, A. (2021). Millennials' evaluation of corporate social responsibility: The wants and needs of the largest and most ethical generation. *Journal of Consumer Behaviour, 20*(3), 521–534.

CrescentRating. (2017). *Global Muslim travel index 2017.* https://newsroom.mastercard.com/asia-pacific/files/2017/05/Report-Mastercard-CrescentRating-GMTI-2017-20mb.pdf

Dabija, D. C., Bejan, B. M., & Dinu, V. (2019). How sustainability oriented is Generation Z in retail? A literature review. *Transformations in Business & Economics, 18*(2).

Dabija, D. C., Bejan, B. M., & Puşcaş, C. (2020). A qualitative approach to the sustainable orientation of generation z in retail: The case of Romania. *Journal of Risk and Financial Management, 13*(7), 152.

D'Arco, M., Marino, V., & Resciniti, R. (2023). Exploring the pro-environmental behavioral intention of Generation Z in the tourism context: The role of injunctive social norms and personal norms. *Journal of Sustainable Tourism.* https://doi.org/10.1080/09669582.2023.2171049

Djafarova, E., & Foots, S. (2022). Exploring ethical consumption of Generation Z: Theory of planned behaviour. *Young Consumers.* https://doi.org/10.1108/YC-10-2021-1405

Dudekula, R. (2023). *Study: Sustainability key in decision making for Muslim Gen Z travellers.* www.marketing-interactive.com/sustainability-key-in-decision-making-for-muslim-gen-z-travellers

Emekli, G., Südaş, I., & Kaba, B. (2023). The travel motivations and experiences of Turkish solo women travellers. In C. M. Hall, S. Seyfi, & S. M. Rasoolimanesh (Eds.), *Contemporary Muslim travel cultures* (pp. 198–214). Routledge.

Gomes, S., Lopes, J. M., & Nogueira, S. (2023). Willingness to pay more for green products: A critical challenge for Gen Z. *Journal of Cleaner Production, 390*(1), 136092.

Green Match. (2020). *4 sustainable behaviours of Gen Z's shopping habits.* www.greenmatch.co.uk/blog/2018/09/gen-zs-sustainable-shopping-habits

Hall, C. M. (2019). Constructing sustainable tourism development: The 2030 agenda and the managerial ecology of sustainable tourism. *Journal of Sustainable Tourism, 27*(7), 1044–1060.

Hall, C. M. (2022). Tourism and the Capitalocene: From green growth to ecocide. *Tourism Planning & Development, 19*(1), 61–74.

Hall, C. M., Mahdavi, M. A., Oh, Y., & Seyfi, S. (2023b). Contemporary Muslim travel and tourism: Cultures and consumption. In C. M. Hall, S. Seyfi, & S. M. Rasoolimanesh (Eds.), *Contemporary Muslim travel cultures* (pp. 14–46). Routledge.

Hall, C. M., & Prayag, G. (Eds.). (2019). *The Routledge handbook of halal hospitality and Islamic tourism.* Routledge.

Hall, C. M., Seyfi, S., & Rasoolimanesh, S. M. (Eds.). (2023a). *Contemporary Muslim travel cultures: Practices, complexities and emerging issues.* Routledge.

Karpovic, D., & Eminens, H. (2018). *SDGs for the Generation Z.* http://socisdg.com/en/blog/sdgs-for-the-generation-z/

Kervankiran, I., Deniz, A., & İlban, K. (2023). The travel pattern and experiences of Turkish female outbound tourists. In C. M. Hall, S. Seyfi, & S. M. Rasoolimanesh (Eds.), *Contemporary Muslim travel cultures* (pp. 183–197). Routledge.

Last, A. (2014). *5 reasons Generation Z could be the ones to save us.* https://sustainablebrands.com/read/stakeholder-trends-and-insights/5-reasons-generation-z-could-be-the-ones-to-save-us

Lipka, M. (2016). *Muslims and Islam: Key findings in the U.S. and around the world.* Pew Research Center.

Liu, J., Wang, C., Zhang, T., & Qiao, H. (2022). Delineating the effects of social media marketing activities on Generation Z travel behaviors. *Journal of Travel Research.* https://doi.org/10.1177/00472875221106394

LoDuca, I. (2020). *Why Gen Z voices matter in making business sustainable.* www.greenbiz.com/article/why-gen-z-voices-matter-making-business-sustainable

Masdar. (2016). *Masdar global Gen Z sustainability survey.* https://masdar.ae/-/media/corporate/downloads/wiser/masdar_gen_z_global_sustainability_survey_top_findings_english.pdf

Mastercard Crescent-Rating. (2019). *Muslim women in travel.* Mastercard Crescent-Rating.

Mastercard Crescent-Rating. (2023). *Gen Z the next generation of travellers.* Mastercard Crescent-Rating.

Nikjoo, A., Markwell, K., Nikbin, M., & Hernández-Lara, A. B. (2021). The flag-bearers of change in a patriarchal Muslim society: Narratives of Iranian solo female travelers on Instagram. *Tourism Management Perspectives, 38,* 100817.

Nonomura, R. (2017). Political consumerism and the participation gap: Are boycotting and "buycotting" youth-based activities? *Journal of Youth Studies, 20*(2), 234–251.

Nugent, C. (2019, November 21). Terrified of climate change? You might have eco-anxiety. *Time.* https://time.com/5735388/climate-change-eco-anxiety/

Oktadiana, H., Pearce, P. L., & Li, J. (2020). Let's travel: Voices from the millennial female Muslim travellers. *International Journal of Tourism Research, 22*(5), 551–563.

Otegui-Carles, A., Araújo-Vila, N., & Fraiz-Brea, J. A. (2022). Solo travel research and its gender perspective: A critical bibliometric review. *Tourism and Hospitality, 3*(3), 733–751.

Plautz, J. (2020, February 3). The environmental burden of Generation Z. *Washington Post.* www.washingtonpost.com/magazine/2020/02/03/eco-anxiety-is-overwhelming-kids-wheres-line-between-education-alarmism/?arc404=true

Polas, M. R. H., Raju, V., Hossen, S. M., Karim, A. M., & Tabash, M. I. (2022). Customer's revisit intention: Empirical evidence on Gen-Z from Bangladesh towards halal restaurants. *Journal of Public Affairs, 22*(3), e2572.

Prayag, G., Aquino, R. S., Hall, C. M., Chen, N., & Fieger, P. (2022). Is Gen Z really that different? Environmental attitudes, travel behaviours and sustainability practices of international tourists to Canterbury, New Zealand. *Journal of Sustainable Tourism, 1–22.* https://doi.org/10.1080/09669582.2022.2131795

Pung, J. M., Yung, R., Khoo-Lattimore, C., & Del Chiappa, G. (2020). Transformative travel experiences and gender: A double duoethnography approach. *Current Issues in Tourism, 23*(5), 538–558.

Rasoolimanesh, S. M., Ramakrishna, S., Hall, C. M., Esfandiar, K., & Seyfi, S. (2020). A systematic scoping review of sustainable tourism indicators in relation to the sustainable development goals. *Journal of Sustainable Tourism.* https://doi.org/10.1080/09669582.2020.1775621

Roberts, S. (2020). *Environmental awareness and tourism consumption: The case of post-millennials* [Unpublished BA Thesis, University of Central Lancashire].

Sandpiper Communications. (2020). *World mental health day: Study highlights brave face of Gen Z in Asia Pacific amid COVID-19.* https://sandpipercomms.com/health/world-mental-health-day-study-highlights-brave-face-of-gen-z-in-asia-pacific-amid-covid-19/

Sebova, L., Pompurova, K., Marcekova, R., & Albertova, A. (2021, June). Solo female travelers as a new trend in tourism destinations. In V. Katsoni & C. van Zyl (Eds.), *Culture and tourism in a smart, globalized, and sustainable world: Springer proceedings in business and economics* (pp. 311–323). Springer.

Seyfi, S., Hall, C. M., Saarinen, J., & Vo-Thanh, T. (2021). Understanding drivers and barriers affecting tourists' engagement in digitally mediated pro-sustainability boycotts. *Journal of Sustainable Tourism.* https://doi.org/10.1080/09669582.2021.2013489

Seyfi, S., Hall, C. M., Vo-Thanh, T., & Zaman, M. (2022). How does digital media engagement influence sustainability-driven political consumerism among Gen Z tourists? *Journal of Sustainable Tourism.* https://doi.org/10.1080/09669582.2022.2112588

Seyfi, S., Sharifi-Tehrani, M., Hall, C. M., & Vo-Thanh, T. (2023). Exploring the drivers of Gen Z tourists' buycott behaviour: A lifestyle politics perspective. *Journal of Sustainable Tourism.* https://doi.org/10.1080/09669582.2023.2166517

Sharpley, R. (2021). On the need for sustainable tourism consumption. *Tourist Studies, 21*(1), 96–107.

Singh, P., Henninger, C. E., Oates, C. J., Newman, N., & Alevizou, P. J. (2022). Children and young people: Opportunities and tensions for sustainability marketing. *Journal of Marketing Management, 38*(9–10), 831–843.

Southan, J. (2019). Solo travel among Gen Z women increases 88 per cent. *Globetrender.* https://globetrender.com/2019/08/07/solo-travel-gen-z-women/

Sudbury Riley, L., Kohlbacher, F., & Hofmeister, A. (2012). A cross-cultural analysis of pro-environmental consumer behaviour among seniors. *Journal of Marketing Management, 28*(3–4), 290–312.

Tayao-Juego, A. (2017). *What Muslim millennial travelers want.* https://business.inquirer.net/240618/muslim-millennial-travelers-want

Vargas-Sanchéz, A., & Perano, M. (2018). Halal tourism through the lens of Generation Z in a Muslim majority country: Implications on tourist services. *International Journal of Business and Management, 13*(9), 36–49.

Yamane, T., & Kaneko, S. (2021). Is the younger generation a driving force toward achieving the sustainable development goals? Survey experiments. *Journal of Cleaner Production, 292*, 125932.

Index

For Product Safety Concerns and Information please contact our EU
representative GPSR@taylorandfrancis.com
Taylor & Francis Verlag GmbH, Kaufingerstraße 24, 80331 München, Germany

www.ingramcontent.com/pod-product-compliance
Lightning Source LLC
Chambersburg PA
CBHW060302220326
41598CB00027B/4200

9 7 8 1 0 3 2 2 6 7 0 7 4